W9-CME-208

HIDDEN SYSTEMS

HIDDEN SYSTEMS

WATER, ELECTRICITY, THE INTERNET, AND THE SECRETS BEHIND THE SYSTEMS WE USE EVERY DAY

DAN NOTT

NEW YORK

HIDDEN SYSTEMS was planned out with pencil on printer paper, drawn with ink on Bristol board, and then lettered and colored digitally.

Visit us on the web! RHKidsGraphic.com • @RHKidsGraphic

Educators and librarians, for a variety of teaching tools, visit us at RHTeachersLibrarians.com

Library of Congress Cataloging-in-Publication Data is available upon request.
ISBN 978-1-9848-9604-9 (paperback) — ISBN 978-0-593-12536-6 (hardcover)
ISBN 978-1-9848-9606-3 (ebook)

Designed by Patrick Crotty

MANUFACTURED IN SINGAPORE
10 9 8 7 6 5 4 3 2
First Edition

A comic on every bookshelf.

For my family—my parents, brother, and grandparents—
for always inspiring kindness, curiosity, and creativity

CONTENTS

Hidden Systems Symbols

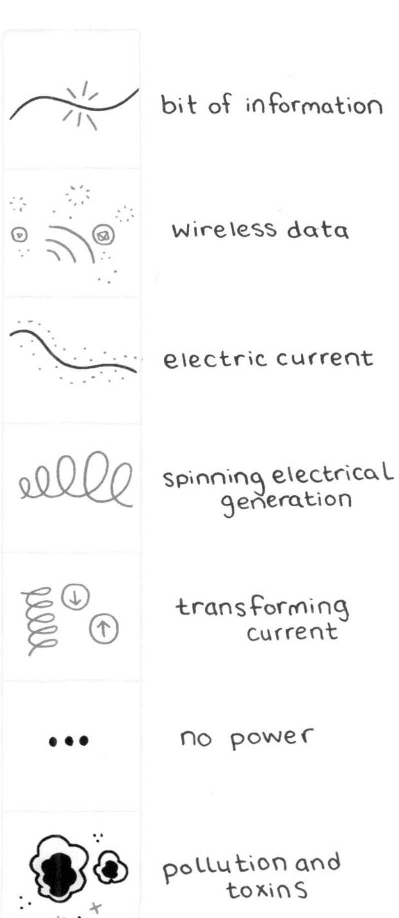

HIDDEN
SYSTEMS

What is a hidden system?

The ultimate, hidden truth of the world
is that it is something that we make,
and could just as easily make differently.

—David Graeber

Sometimes, I am very surprised by how little I know about everyday things.
1924
how is there enough water for everyone all the time?
shhhh
and what's really powering this light?
what even IS electricity?
Usually I just move on.
oh well
sip
But sometimes, a question sticks,
and I'll look it up.
dear google
how does information move around the internet?
Answers are often basic,
through wires, lol
scroll scroll
or technical and hard to visualize,
so I go on not knowing,
and forget about it until something goes very wrong.

A hidden system is something we don't notice
until it breaks.
NO MORE GAS
SRY
...
...
...
huh. Nothing works.
But when these systems are doing what they're supposed to,
they become so commonplace
that we hardly see them.

Hidden systems are in the news all the time.
lead in drinking water
cyberattack hits key U.S. pipeline
levees burst
massive power outages
Usually when something **dramatic** happens.
(especially if something explodes)
But by overlooking hidden systems the **rest of the time,**
we take for granted the **benefits** they provide for **some of us,**
WRRRRR
and disregard the **harm** they cause **others.**
WRRRR
These systems **structure** our society,
and even when **they're working,**
are a source of **inequality**
and **environmental harm.**

I began drawing about hidden systems because comics seem to have this superpower-like ability
to compare how we think about something
the cloud
with how it works concretely.
Bzzzzz
To view small pieces
in the context
of larger systems.
And to show the embedded histories,
inequities,
power disparities
colonial legacies
access
and imagined potentials

all contained
within the things
we **notice least.**

To make this book, I talked with a lot of people and read as much as I could.
books by engineers
and historians
personal experiences
I believe that **asking questions** and **drawing out answers** can be a powerful way of **learning** and **understanding.**
cartoonist
These comics are about the process of forming questions—
how do we picture the internet vs. what is it actually?
how do we power our world with electricity?
how do our water systems interact with the Earth's changing climate?
and discovering a **small piece**
of what's **hidden beneath the surface.**

The network is the computer.

—John Gage, Sun Microsystems

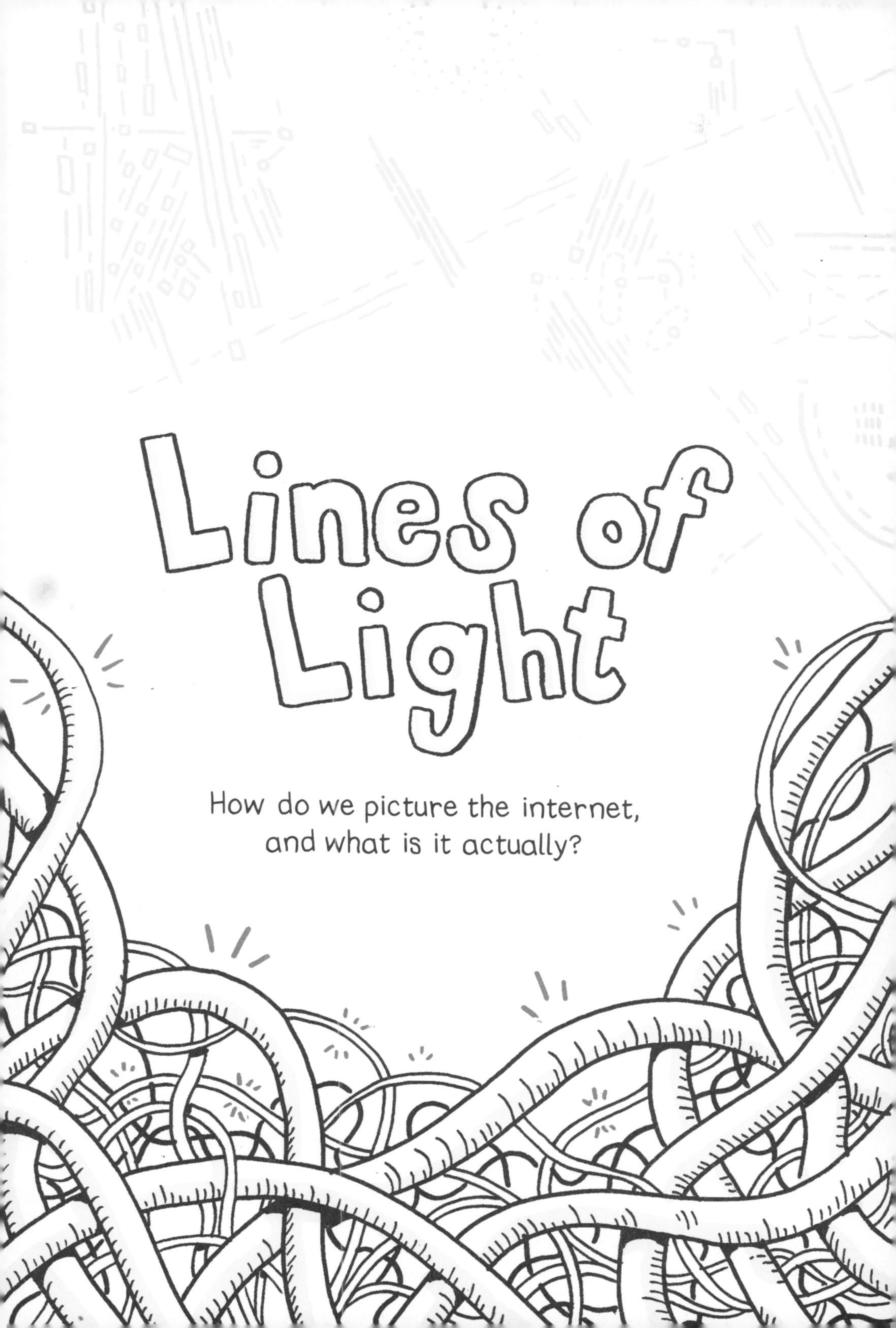

Lines of Light

How do we picture the internet,
and what is it actually?

Sen. Ted Stevens
Net neutrality hearing
2006
They want to deliver
vaaast amounts
of information
over the internet
And again
the internet,
is not something
you just dump
something on,
it's not a
big truck,
it's — it's
a series of tubes!

The internet can seem pretty abstract

Sup?

in the way it appears to disregard **time** and **space**.

nm, u?

Maybe that's why we tend to **think** and **talk** about the internet using a **strange mix of visual metaphors.**

But if **metaphors** are **all** we have to talk about the **internet,**

how does that affect how we **picture** and **understand** it?

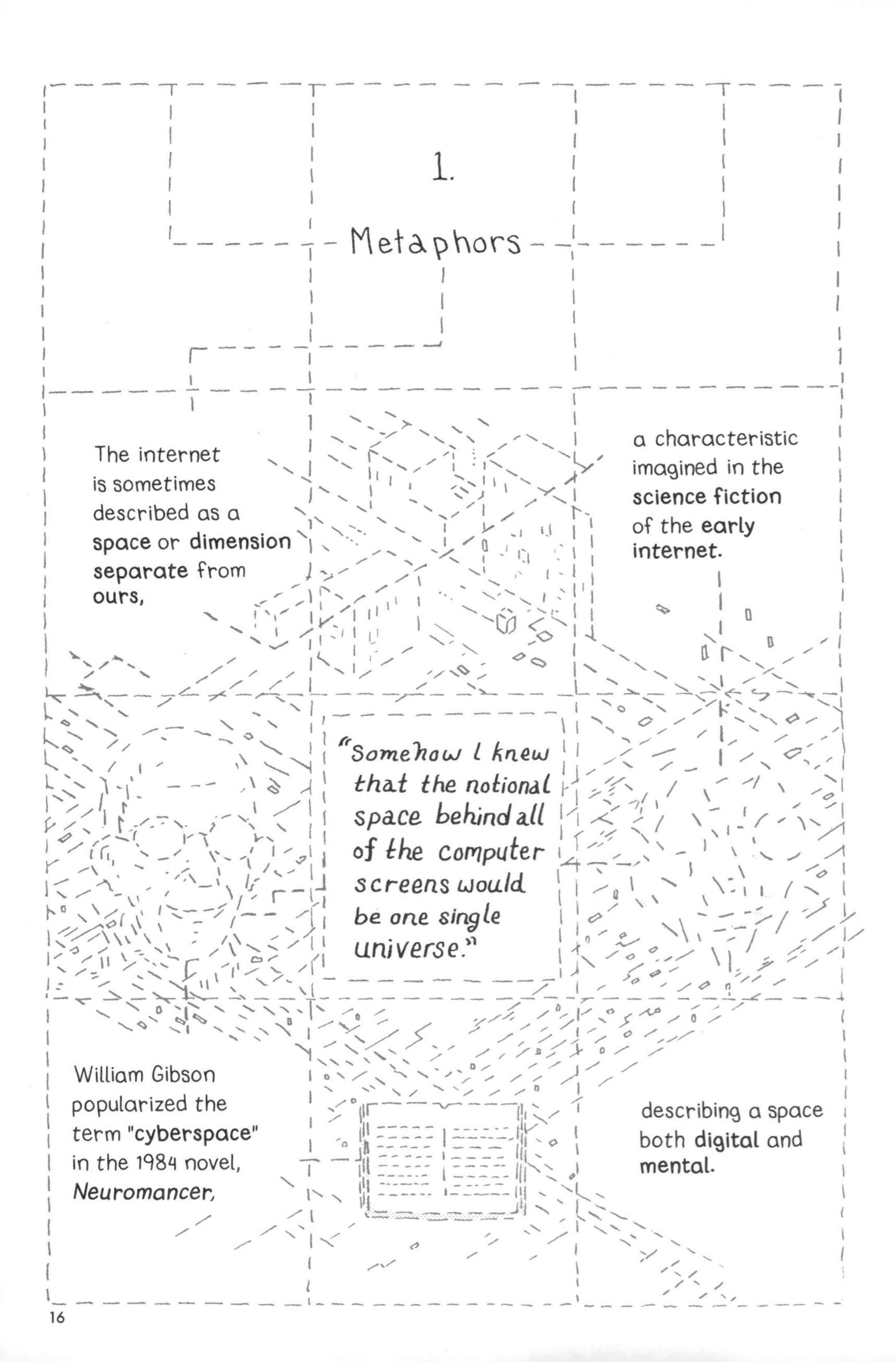
1.
Metaphors
The internet is sometimes described as a **space** or **dimension** **separate** from **ours,**
a characteristic imagined in the **science fiction** of the **early internet.**
"Somehow I knew that the notional space behind all of the computer screens would be one single universe."
William Gibson popularized the term "**cyberspace**" in the 1984 novel, *Neuromancer,*
describing a space both **digital** and **mental.**

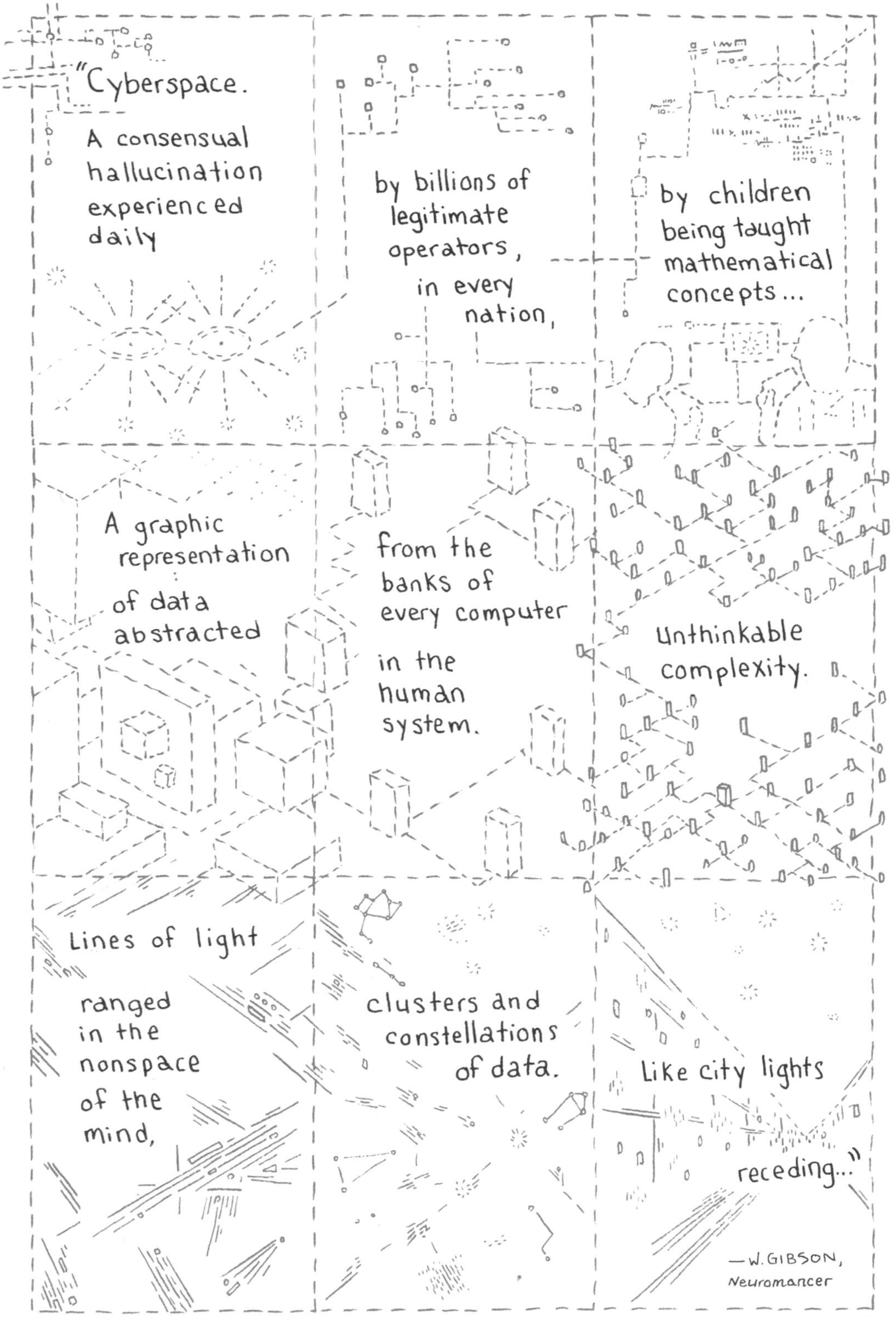

"Cyberspace.

A consensual hallucination experienced daily

by billions of legitimate operators, in every nation,

by children being taught mathematical concepts...

A graphic representation of data abstracted

from the banks of every computer in the human system.

Unthinkable complexity.

Lines of light ranged in the nonspace of the mind,

clusters and constellations of data.

Like city lights receding..."

—W. GIBSON, *Neuromancer*

The **metaphors** we use often reflect our **bias.**
For example, when described as a **transportation infrastructure,**
THE INFORMATION SUPERHIGHWAY
with **on-ramps**
and **online traffic,**
fast lanes and **slow lanes,**
the internet becomes a system to be **built, maintained,** and **regulated** for **public use.**

When we talk about using the internet to **connect and communicate,**

we refer to **"town squares"** or a collection of **"virtual communities."**

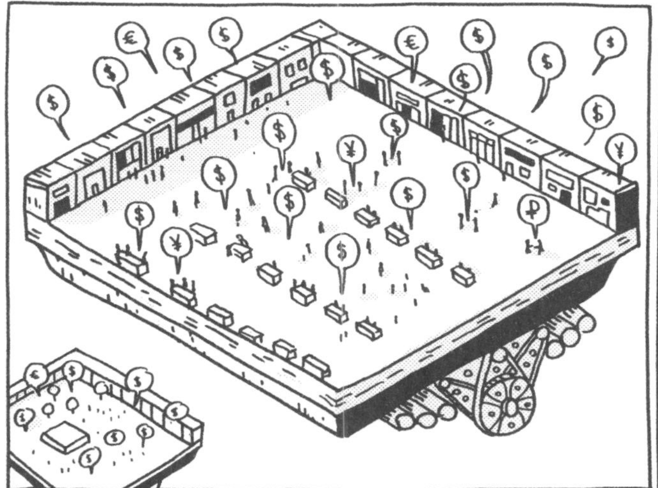

With an **economic** agenda,

the internet is a **marketplace** of commerce and ideas, or an **engine** of growth.

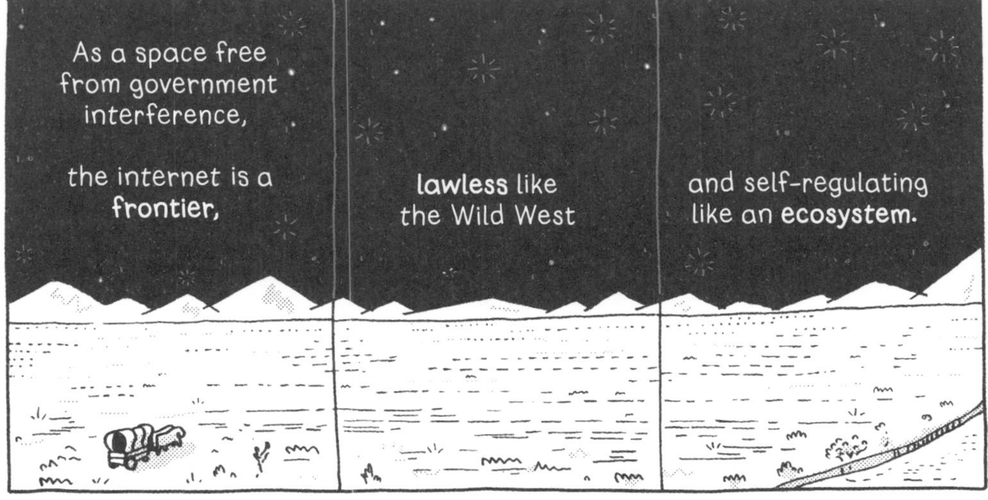

Sometimes, this frontier is described as an **ocean of data,**
with a **surface web**
and a **deep web.**
It's navigated by **internet explorers**
pillaged by **pirates,**
and surfed by the **aimlessly curious.**

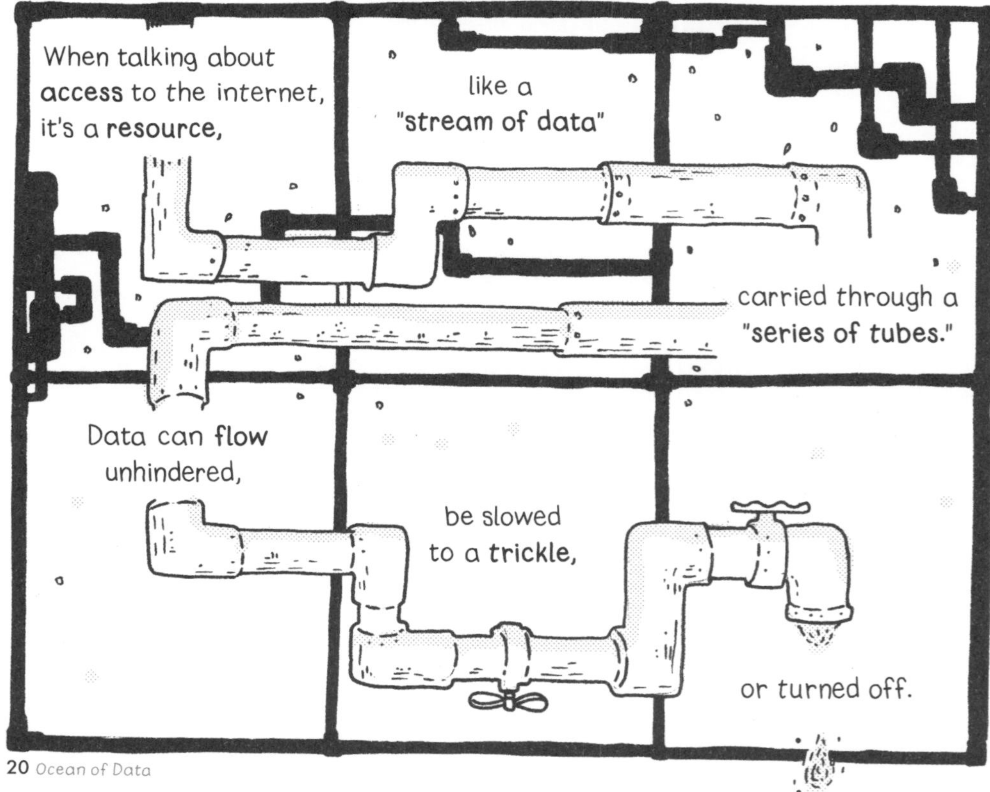

When talking about **access** to the internet, it's a **resource,**
like a **"stream of data"**
carried through a **"series of tubes."**
Data can **flow** unhindered,
be slowed to a **trickle,**
or turned off.

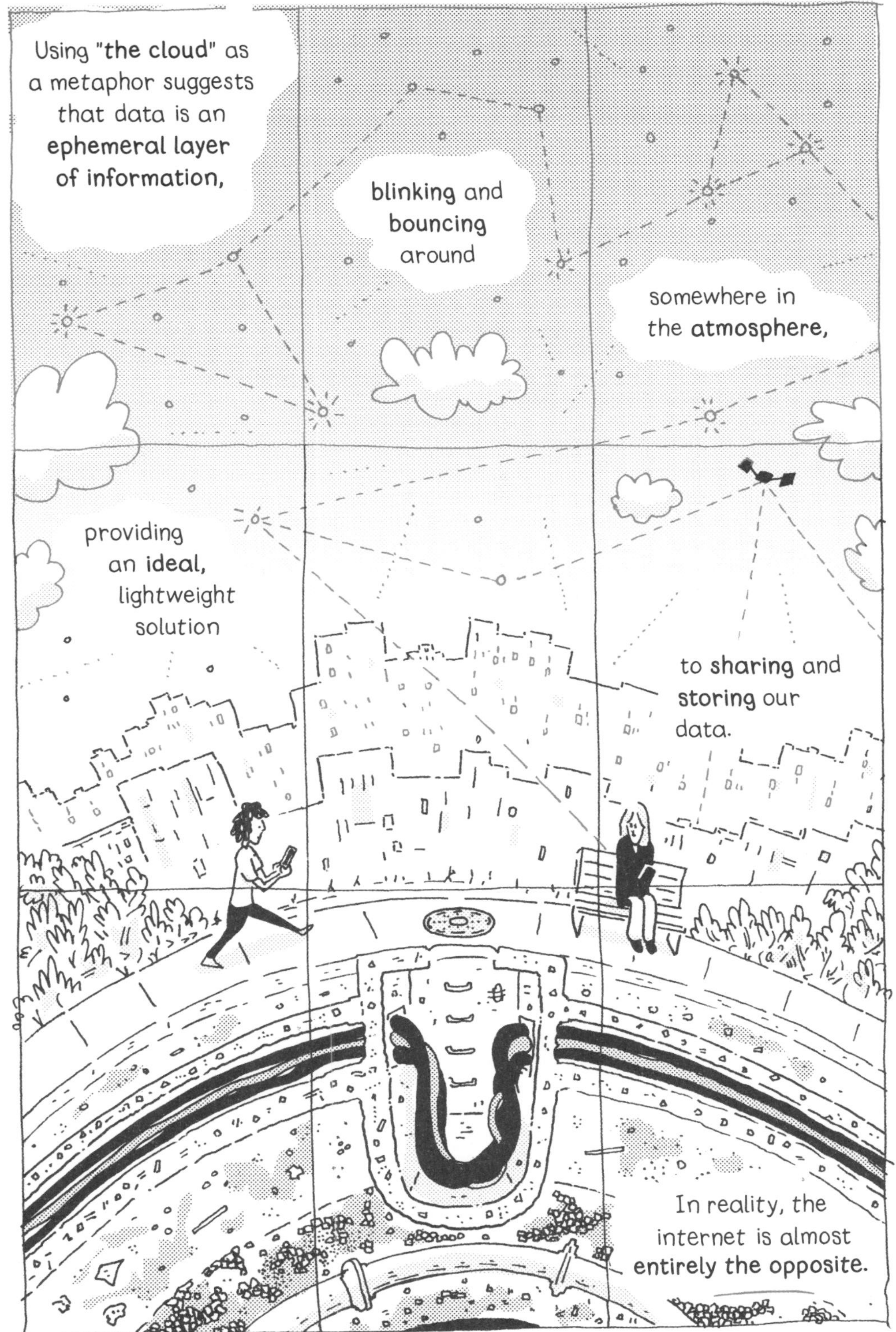
Using "the cloud" as a metaphor suggests that data is an ephemeral layer of information,
blinking and bouncing around
somewhere in the atmosphere,
providing an ideal, lightweight solution
to sharing and storing our data.
In reality, the internet is almost entirely the opposite.

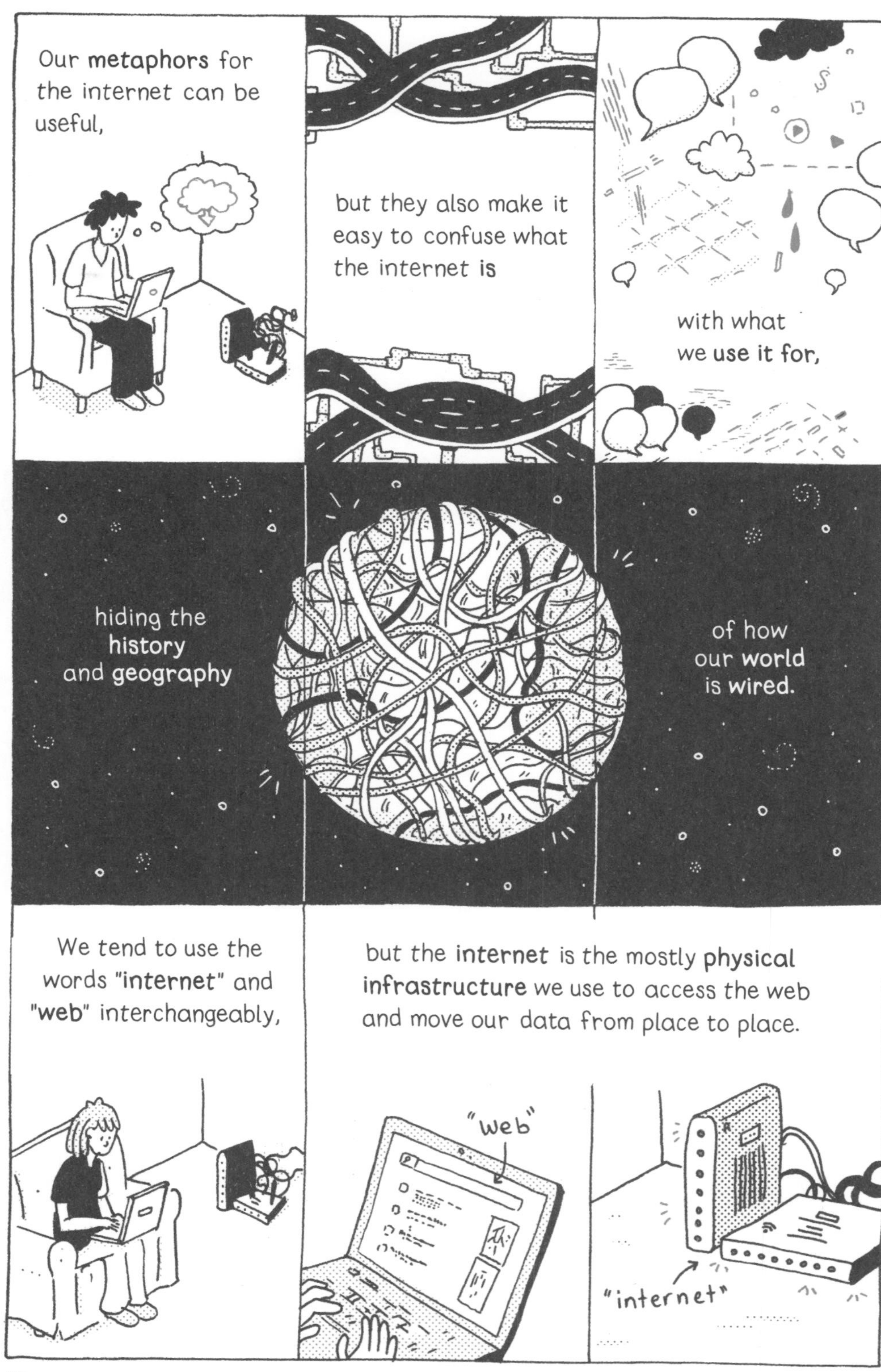
Our **metaphors** for the internet can be useful,
but they also make it easy to confuse what the internet **is**
with what we **use it for**,
hiding the **history** and **geography**
of how our **world** is **wired**.
We tend to use the words "**internet**" and "**web**" interchangeably,
but the **internet** is the mostly **physical infrastructure** we use to access the web and move our data from place to place.
"web"
"internet"

Every day, we're faced with issues about how the internet **works,**
Net neutrality changes could alter the basic premise of the internet.
how it affects us **personally**
Recent breaches by hackers leave consumers concerned about the safety of their data.
NEWS
and as a **society.**
Experts warn that an advanced cyberattack could be catastrophic.
But to **understand** the **problems** with the internet,
we should understand **what it really is,**
where it **came from,**
and what's secretly happening every time we **post, read,** or **watch** something online.

2.
Cables
The way we talk about the internet makes it seem like it **travels through the air.**
cable-laying ship,
pacific ocean

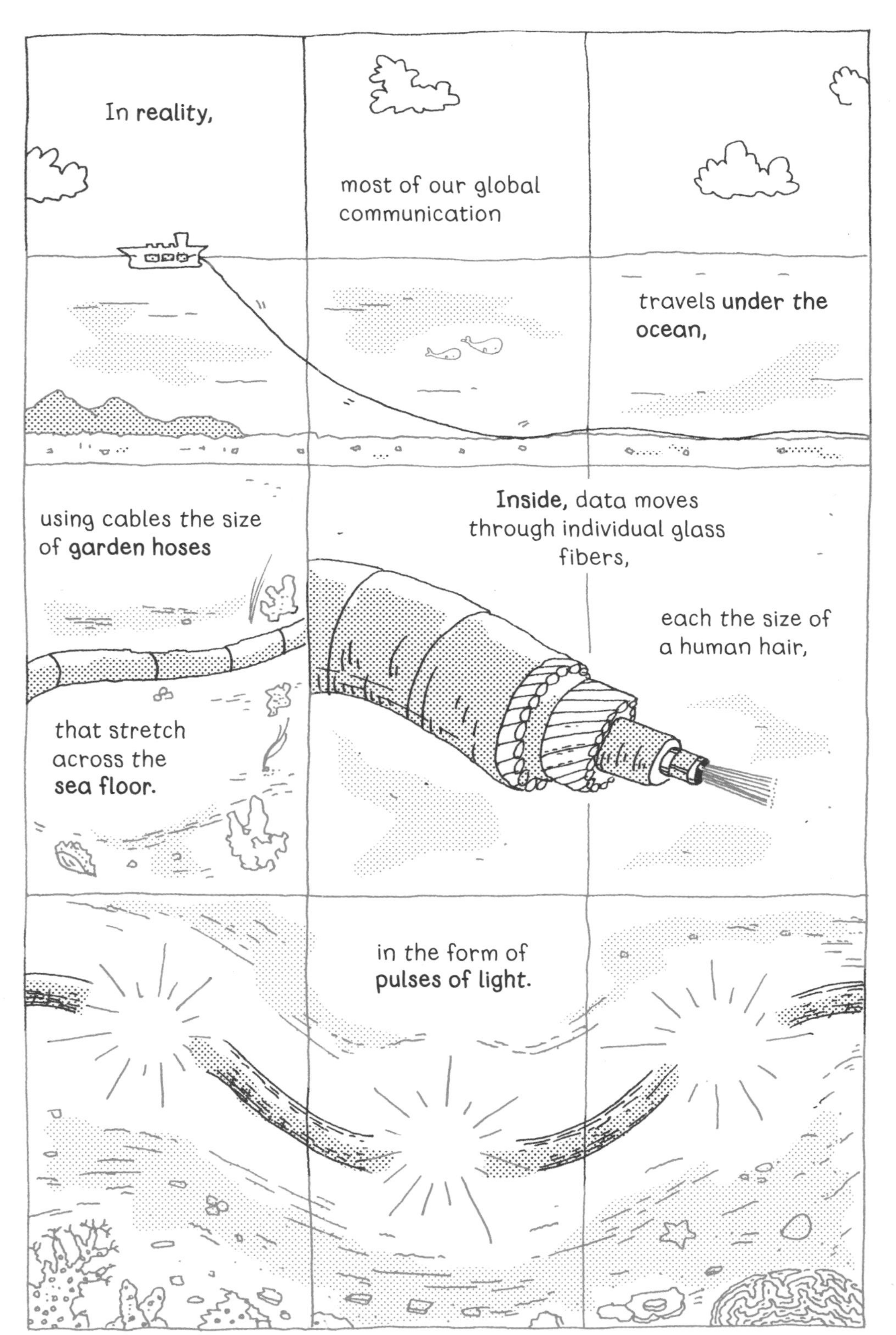
In reality,
most of our global communication
travels under the ocean,
using cables the size of garden hoses
that stretch across the sea floor.
Inside, data moves through individual glass fibers,
each the size of a human hair,
in the form of pulses of light.

As high tech as it seems, these modern cables are built on the principles of our **oldest communications technologies.**

In 1844, Samuel Morse sent the **first telegram using his code** from D.C. to Baltimore,

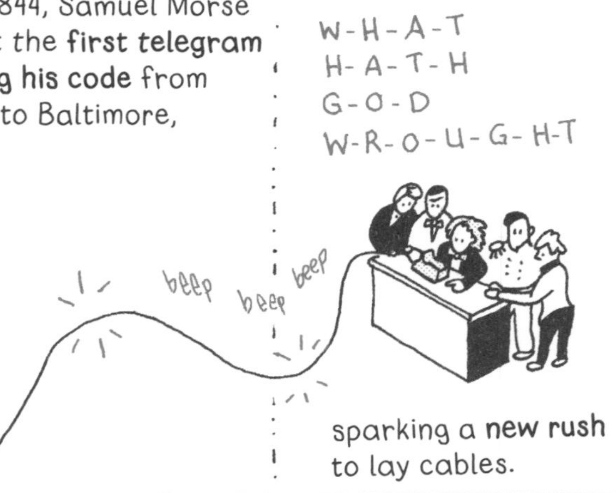

sparking a **new rush** to lay cables.

Soon after, British engineers created a cable that could lay at the bottom of the ocean.

They outfitted ships to lay **thousands of miles** of electric cable

and used it to communicate across their **far-flung empire,**

eventually linking its colonial territories across the planet in a network known as the **"All Red Line."**

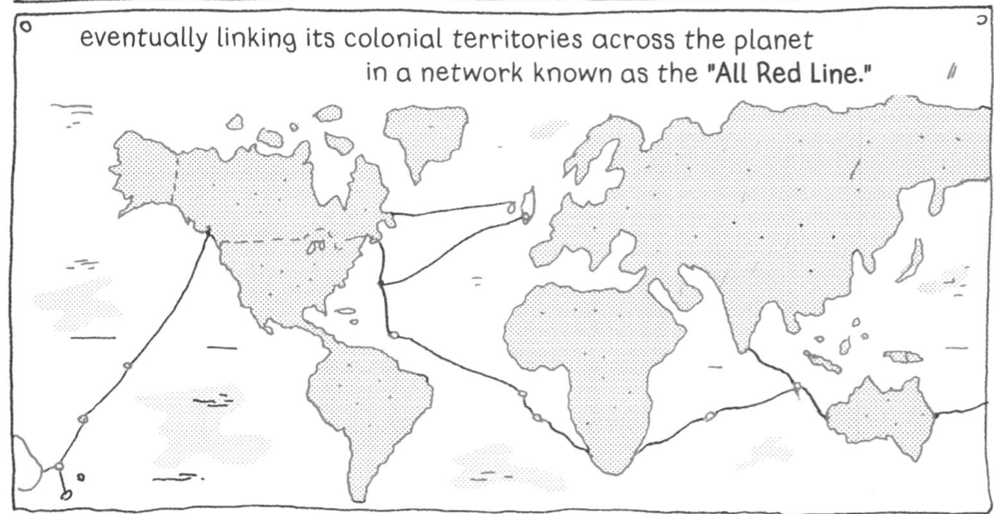

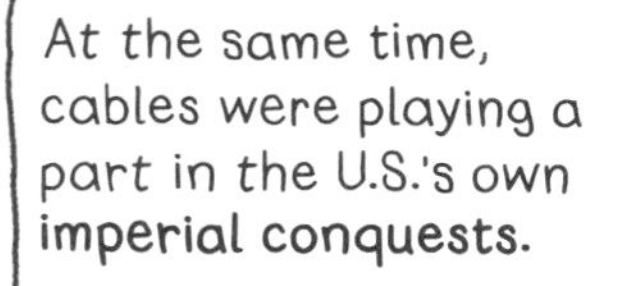

At the same time, cables were playing a part in the U.S.'s own **imperial conquests.**

When the **U.S. attacked Cuba in 1898,** the navy **severed** some of the island's undersea cables in an attempt to cut it off from the **Spanish Empire.**

The U.S. had its **first transpacific cable** built in **1903,** allowing it to communicate with the **Philippines,** which it had recently colonized.

Theodore Roosevelt used the Pacific cable and others to send the **U.S.'s first messages around the world.**

I open the American Pacific cable with greetings to you and the people of the Philippines.

July 4, 1903

The **telegraph network**, through a combination of **imperialism** and a **commercial undersea cable boom,** became the first **globalized communication system** in human history.

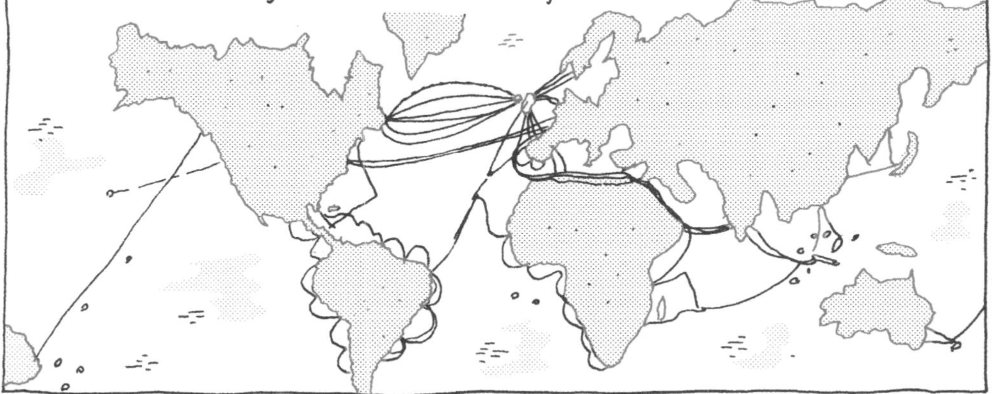

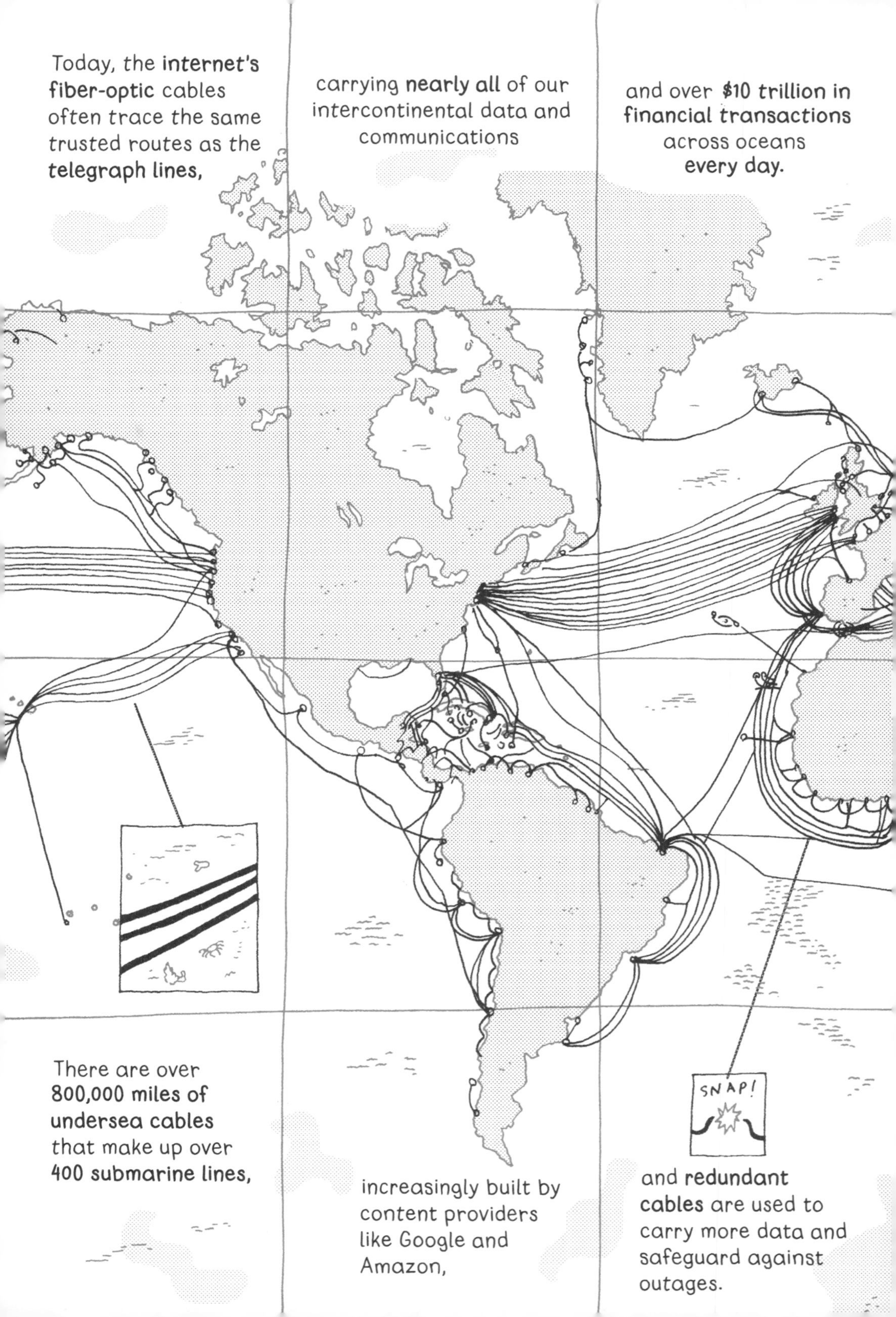
Today, the internet's fiber-optic cables often trace the same trusted routes as the telegraph lines,
carrying nearly all of our intercontinental data and communications
and over $10 trillion in financial transactions across oceans every day.
There are over 800,000 miles of undersea cables that make up over 400 submarine lines,
increasingly built by content providers like Google and Amazon,
SNAP!
and redundant cables are used to carry more data and safeguard against outages.

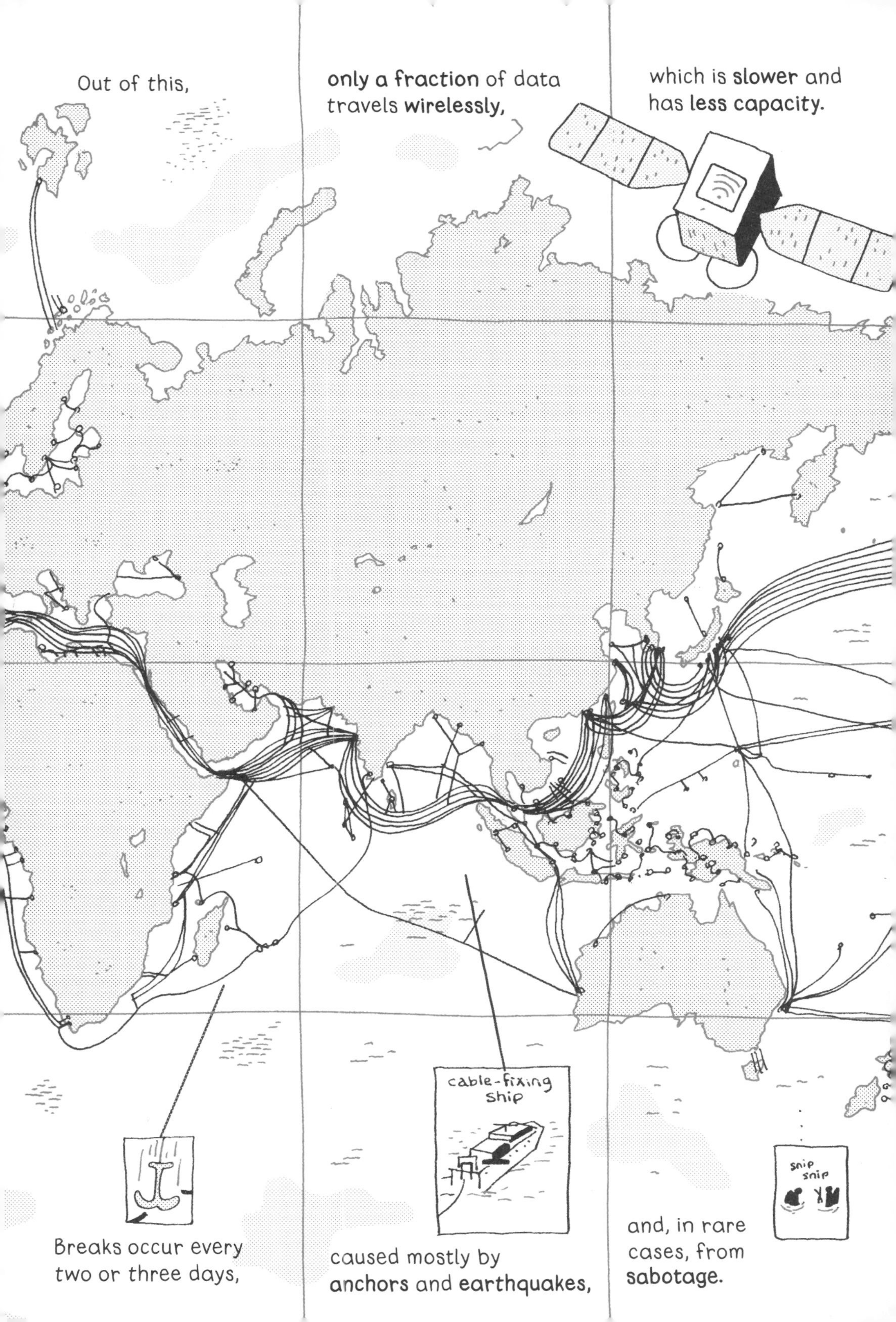
Out of this,
only a fraction of data travels **wirelessly**,
which is **slower** and has **less capacity**.
cable-fixing ship
Breaks occur every two or three days,
caused mostly by **anchors** and **earthquakes**,
snip snip
and, in rare cases, from **sabotage**.

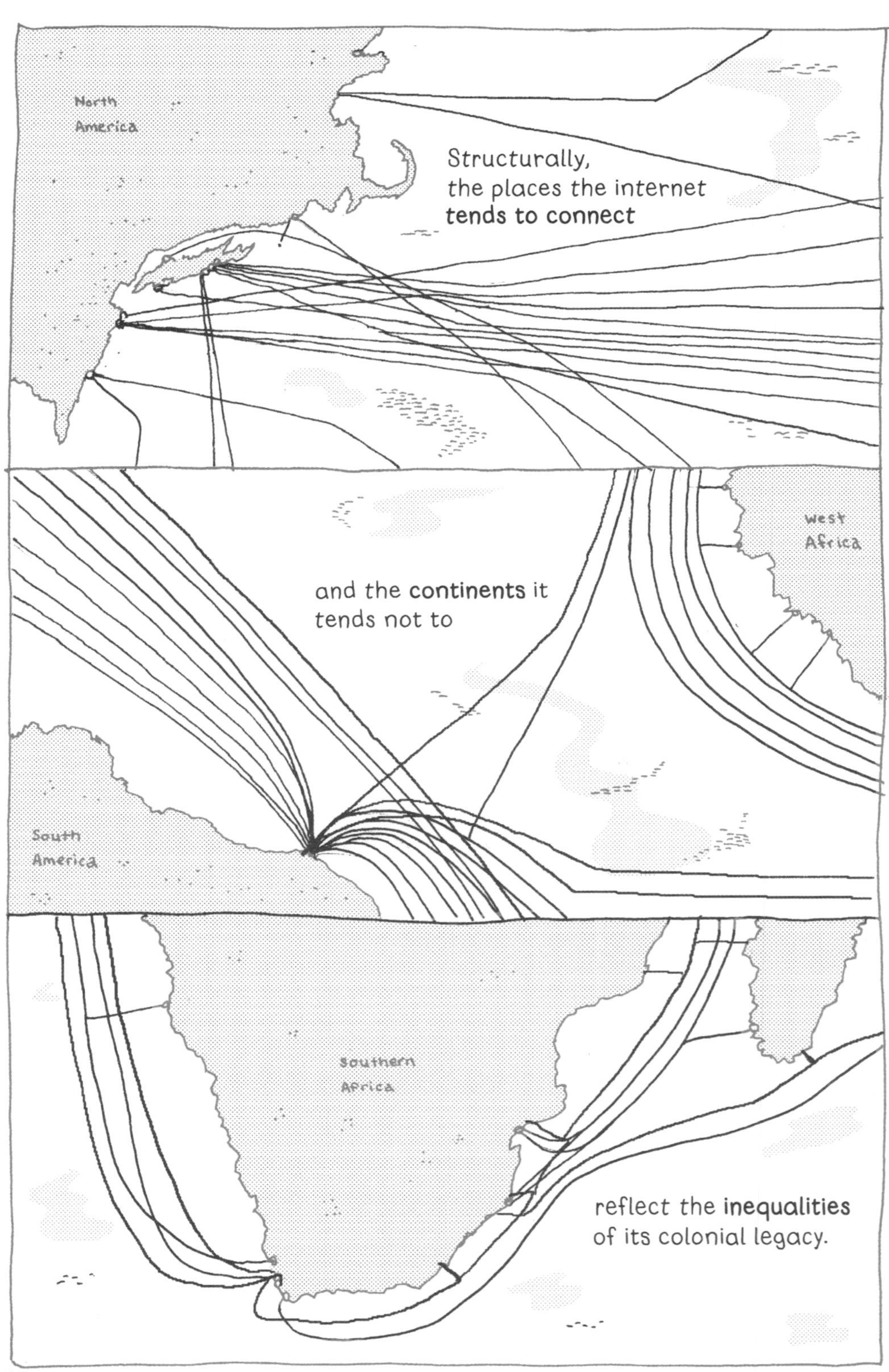
North
America
Structurally,
the places the internet
tends to connect
and the continents it
tends not to
West
Africa
South
America
Southern
Africa
reflect the inequalities
of its colonial legacy.

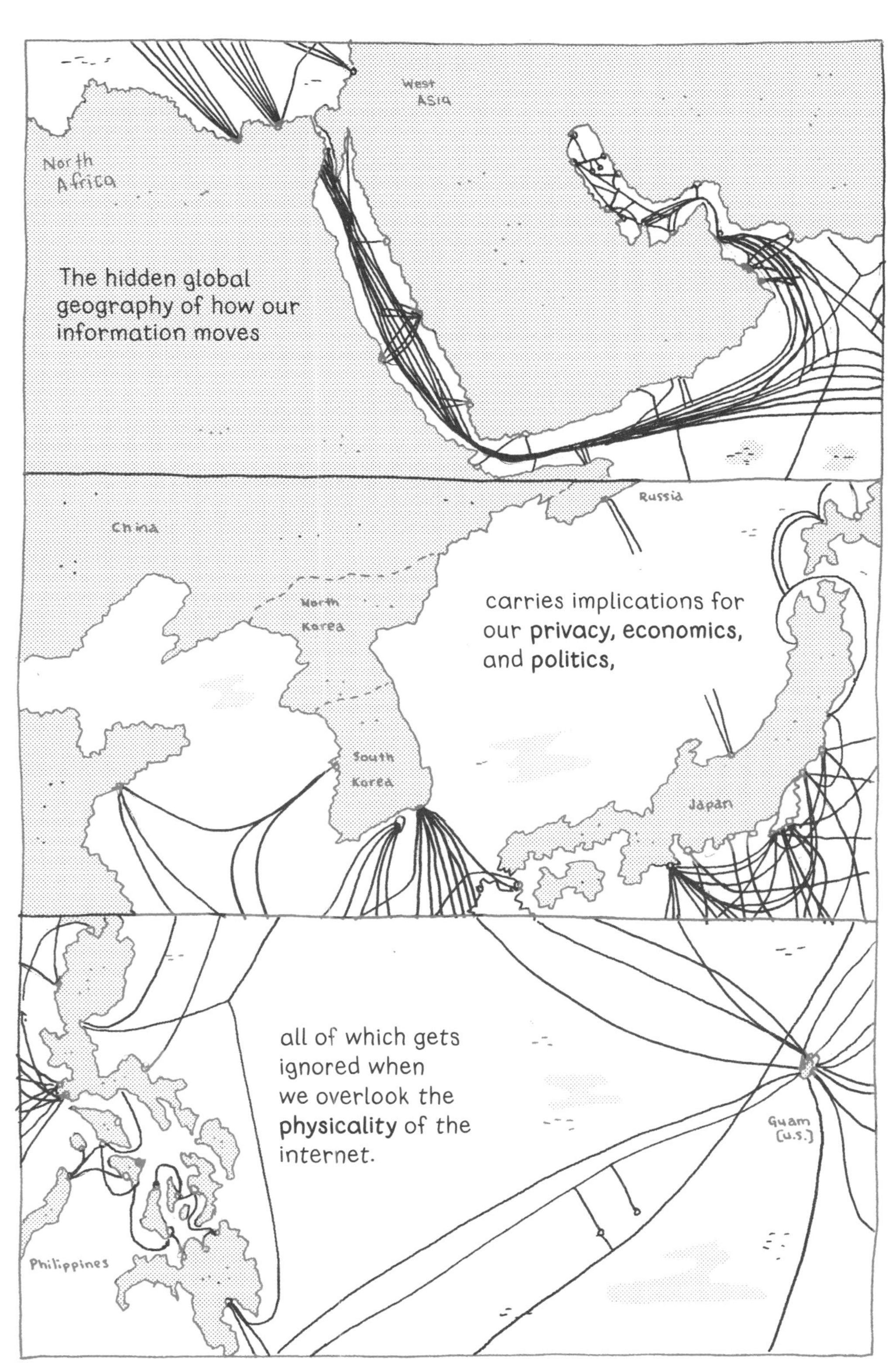
West
Asia
North
Africa
The hidden global
geography of how our
information moves
Russia
China
North
Korea
carries implications for
our **privacy, economics,**
and **politics,**
South
Korea
Japan
all of which gets
ignored when
we overlook the
physicality of the
internet.
Guam
(U.S.)
Philippines

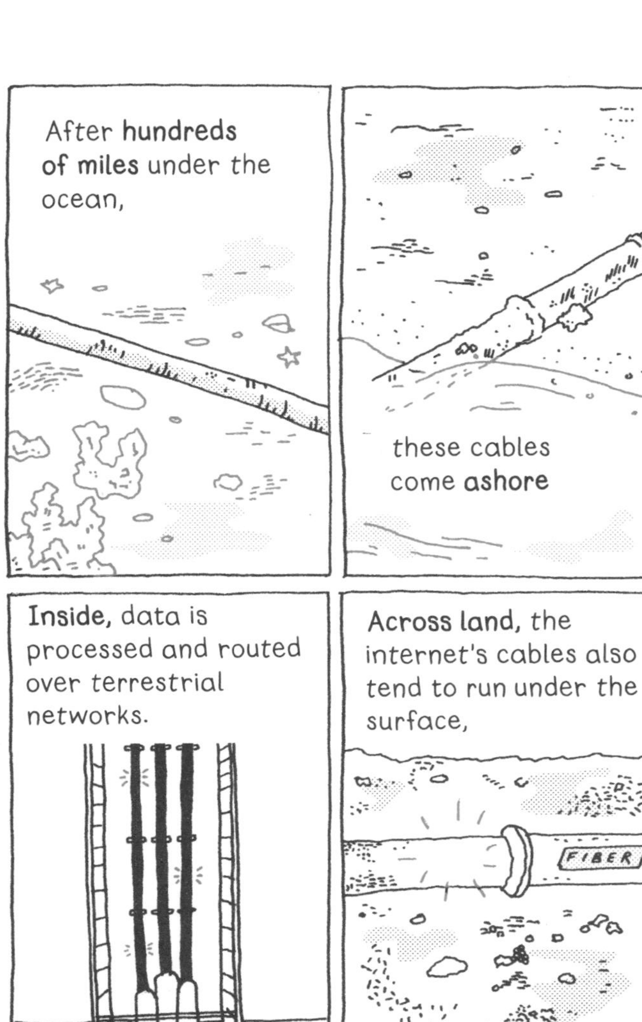
After **hundreds of miles** under the ocean,

these cables come **ashore**

and **connect** to stations not far from the beach.

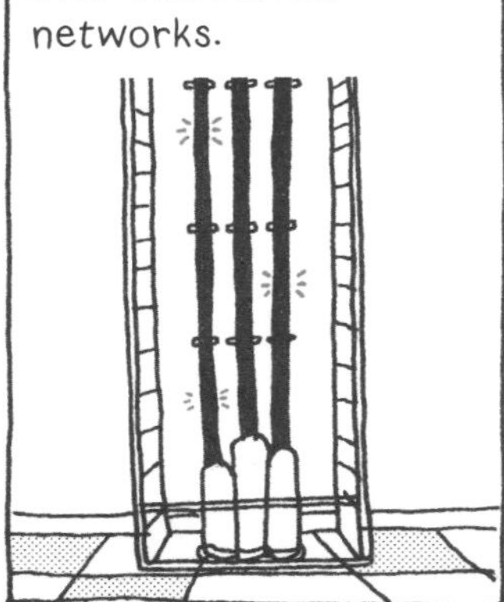
Inside, data is processed and routed over terrestrial networks.

Across land, the internet's cables also tend to run under the surface,
FIBER

buried along **railroad tracks,**

and along major highways where the **rights-of-way** are easier to secure.

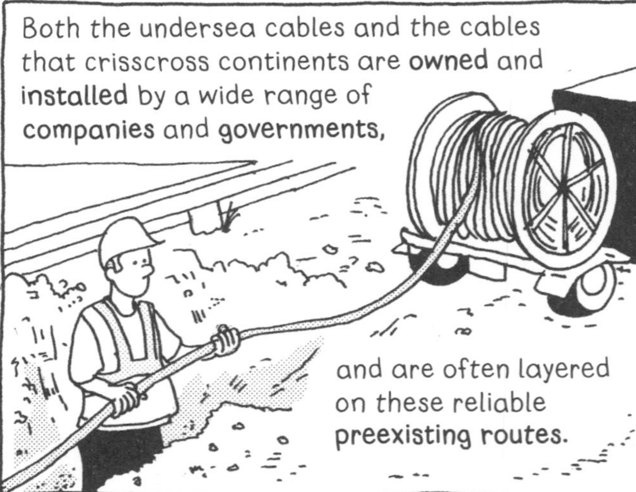
Both the undersea cables and the cables that crisscross continents are **owned** and **installed** by a wide range of **companies** and **governments,**
and are often layered on these reliable **preexisting routes.**

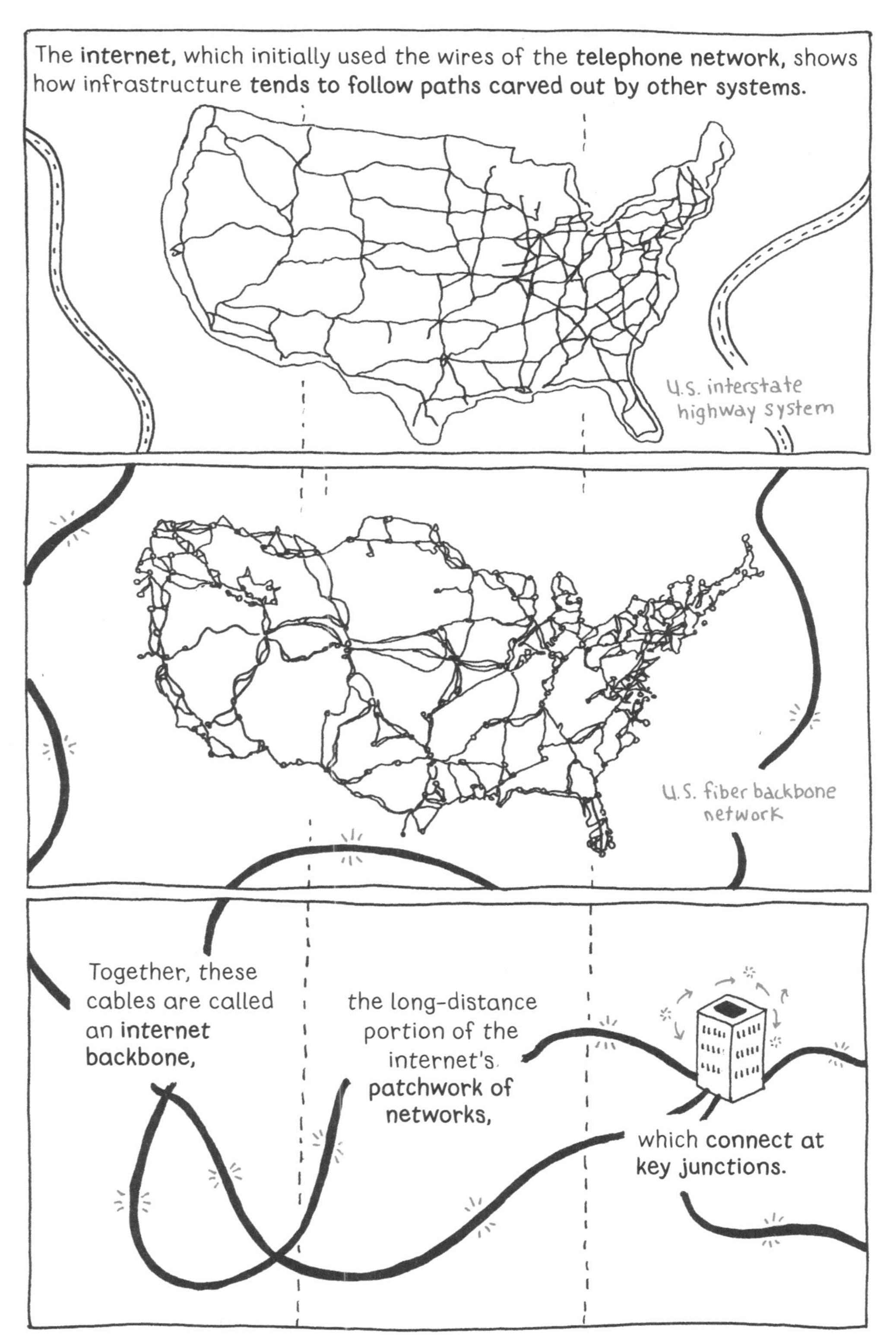
The **internet,** which initially used the wires of the **telephone network,** shows how infrastructure **tends to follow paths carved out by other systems.**
U.S. interstate highway system
U.S. fiber backbone network
Together, these cables are called an **internet backbone,**
the long-distance portion of the internet's **patchwork of networks,**
which **connect at key junctions.**

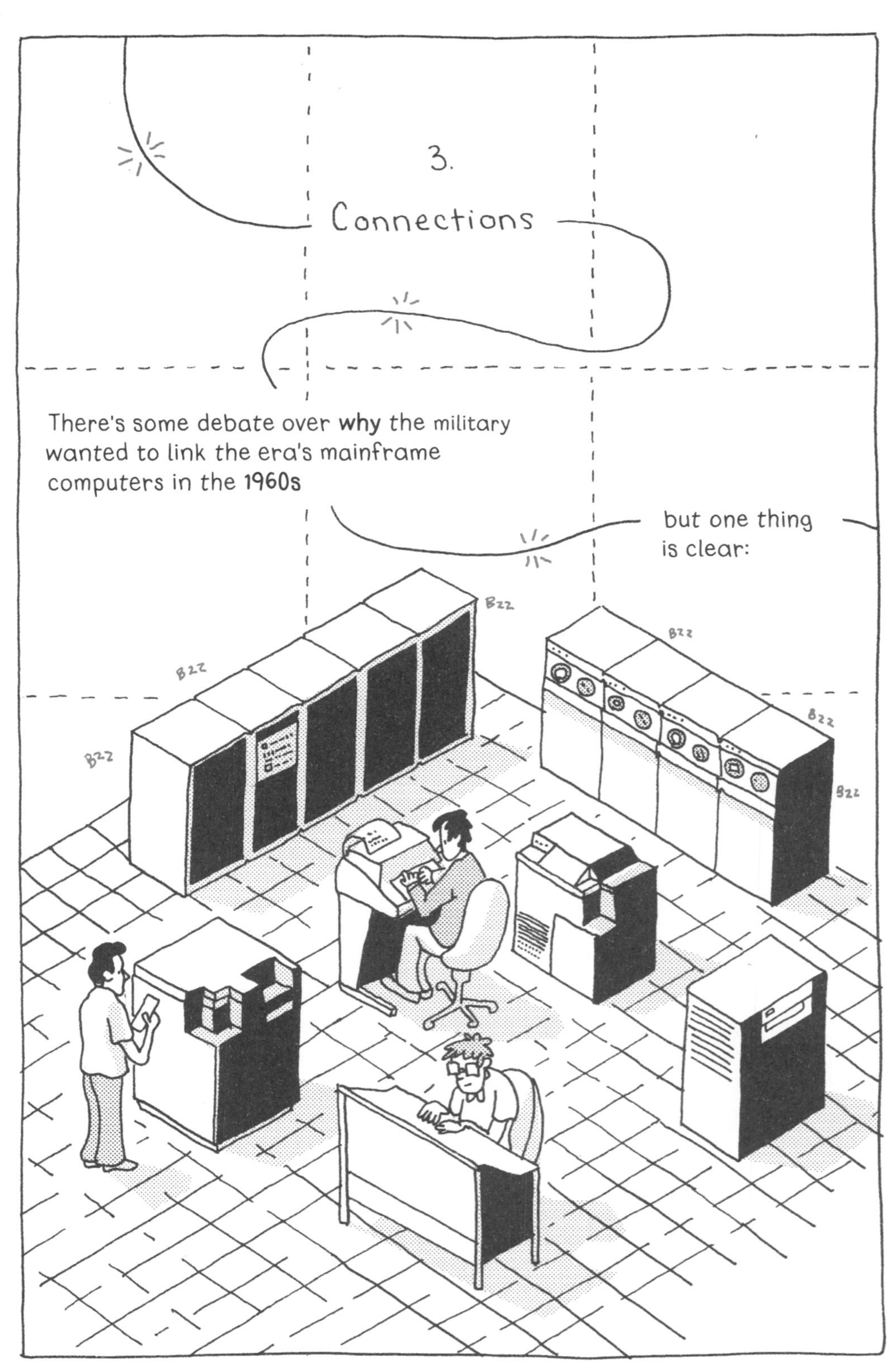
3.
Connections
There's some debate over **why** the military wanted to link the era's mainframe computers in the **1960s**
but one thing is clear:
Bzz
Bzz
Bzz
Bzz
Bzz
Bzz

when UCLA students and staff prepared to **connect two of these room-sized computers** on October 29, 1969,
they had no way of knowing what they were helping to invent.
The lab was working on an experimental computer network for the military's **Advanced Research Projects Agency (ARPA).**
U.S. GOVERNMENT PROPERTY
DAHC-0179-13-
Using new theories of **transmitting data**
Packet Switching
and a futuristic machine called an **IMP**—an **early router** known then as a gateway—
they wanted to connect **two computers** miles apart through the telephone network.
U.S.A
It **worked,**
but the system **crashed** from memory overload halfway through exchanging the words "**log-in.**"
The first message sent over the network was "**L-O.**"

As more routers were installed
UCLA
and connected to mainframe computers around the country,
SRI
the size of the network, called the ARPANET, expanded using the wires of the telephone network,
1969
UTAH
UCSB
linking together
U.S. military and academic institutions
from coast to coast.
STANFORD
SDC
RAND
1970
LINCOLN
MIT
CASE
BBN
HARVARD
CARNEGIE
The ARPANET was used by academics for research
and also, shortly after being built,
1973
by the military to help manage the surveillance of Americans.

The **ARPANET** only had nodes with the military and with **universities that took ARPA funding.**

The researchers working in these departments helped to write the **open protocols** for this early network.

Email, which was invented on a **whim** and was a **surprise success,** is still largely **unchanged** today.

Companies like **IBM** and **Xerox** saw the use in linking their computers

and, along with agencies like **NASA,** created their own distinct networks—

but without a **common networking language** between them, they couldn't communicate.

In 1983, a protocol called **TCP/IP** was mandated, creating **universal rules** for how data was **sent, routed, and received,**

allowing a **scattering** of **separate networks** to **interconnect**

and take the form of the early **internet.**

In the **1980s**, the internet's uses were still basic— **file transfers, email, chat rooms**—

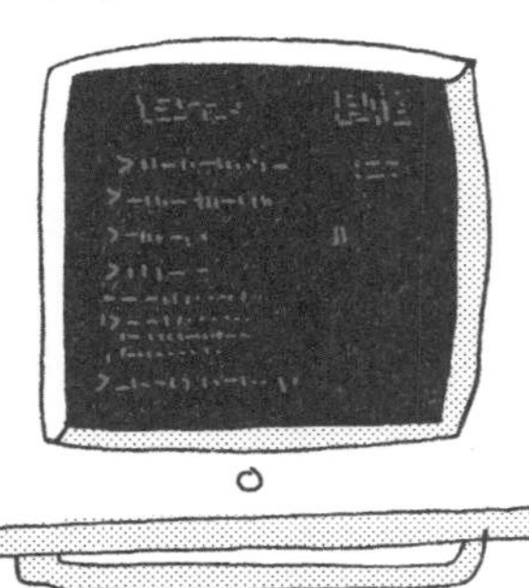

but in 1991, the **World Wide Web** was released,

and early **visual browsers** like **Mosaic** and **Netscape** created new demand for internet access.

At the time, the **National Science Foundation (NSF)** was in charge of the internet's main backbone,

and it didn't allow **commercial traffic** on its national network.

A few early internet providers had begun laying their own fiber-optic cables,

but in order to exchange commercial traffic between their separate networks,

they needed somewhere to **physically connect them.**

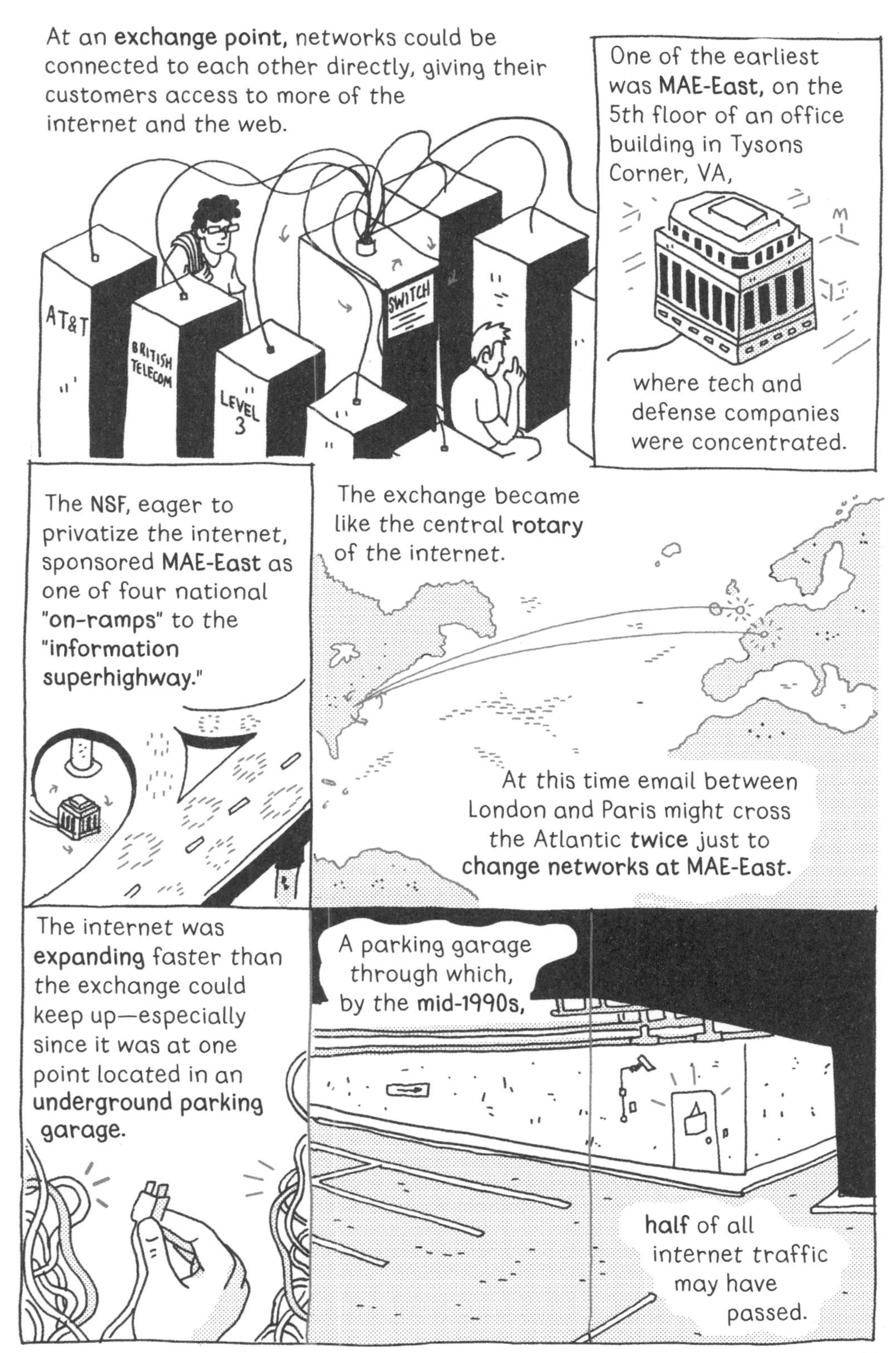
At an **exchange point,** networks could be connected to each other directly, giving their customers access to more of the internet and the web.
AT&T
BRITISH TELECOM
LEVEL 3
SWITCH
One of the earliest was **MAE-East,** on the 5th floor of an office building in Tysons Corner, VA,
where tech and defense companies were concentrated.
The **NSF,** eager to privatize the internet, sponsored **MAE-East** as one of four national **"on-ramps"** to the **"information superhighway."**
The exchange became like the central **rotary** of the internet.
At this time email between London and Paris might cross the Atlantic **twice** just to **change networks at MAE-East.**
The internet was **expanding** faster than the exchange could keep up—especially since it was at one point located in an **underground parking garage.**
A parking garage through which, by the **mid-1990s,**
half of all internet traffic may have passed.

Today, **exchange points** are spread out across the world,

but are **concentrated** in the **U.S. and Western Europe,**

where most of the **content** on the internet is stored.

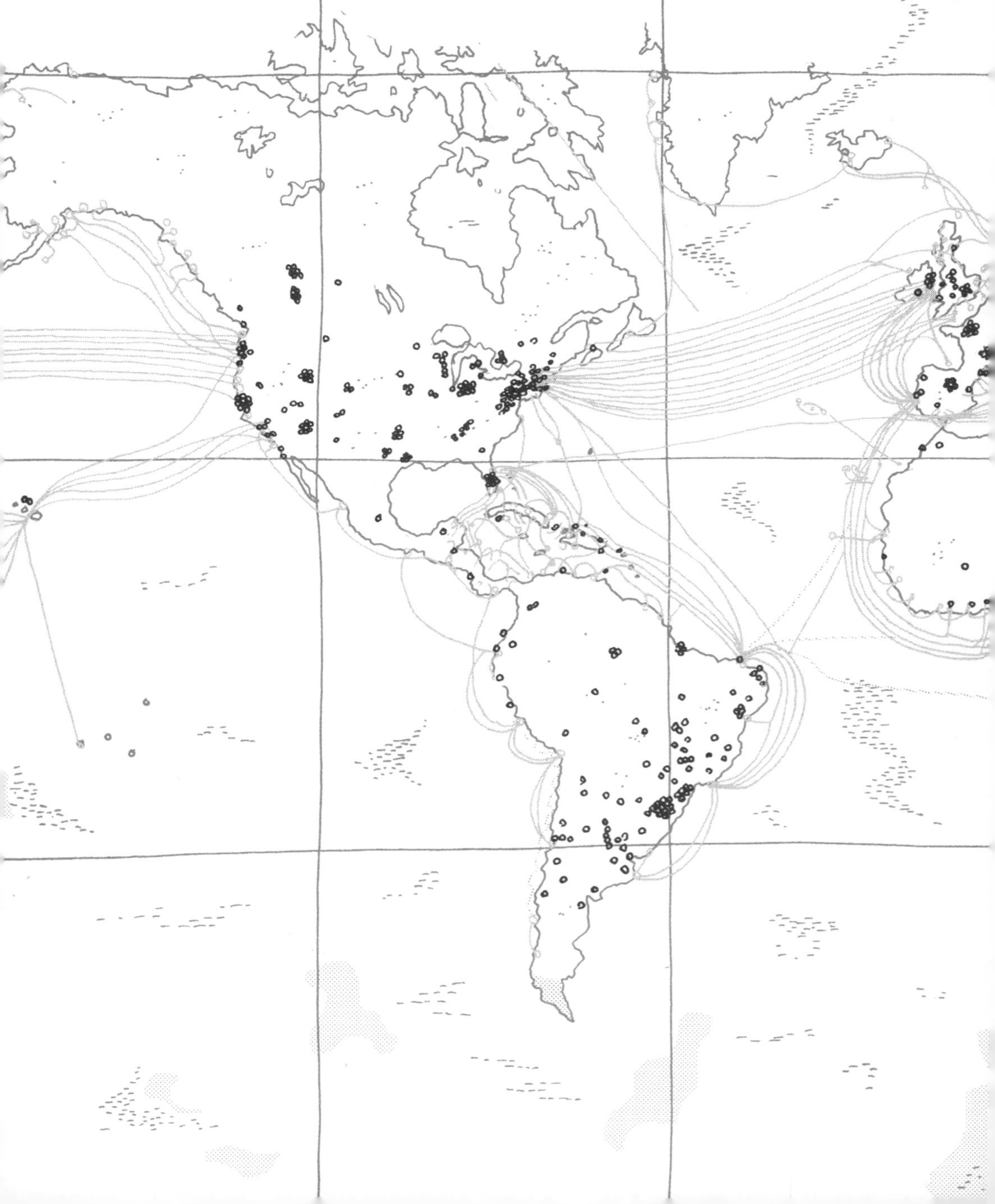

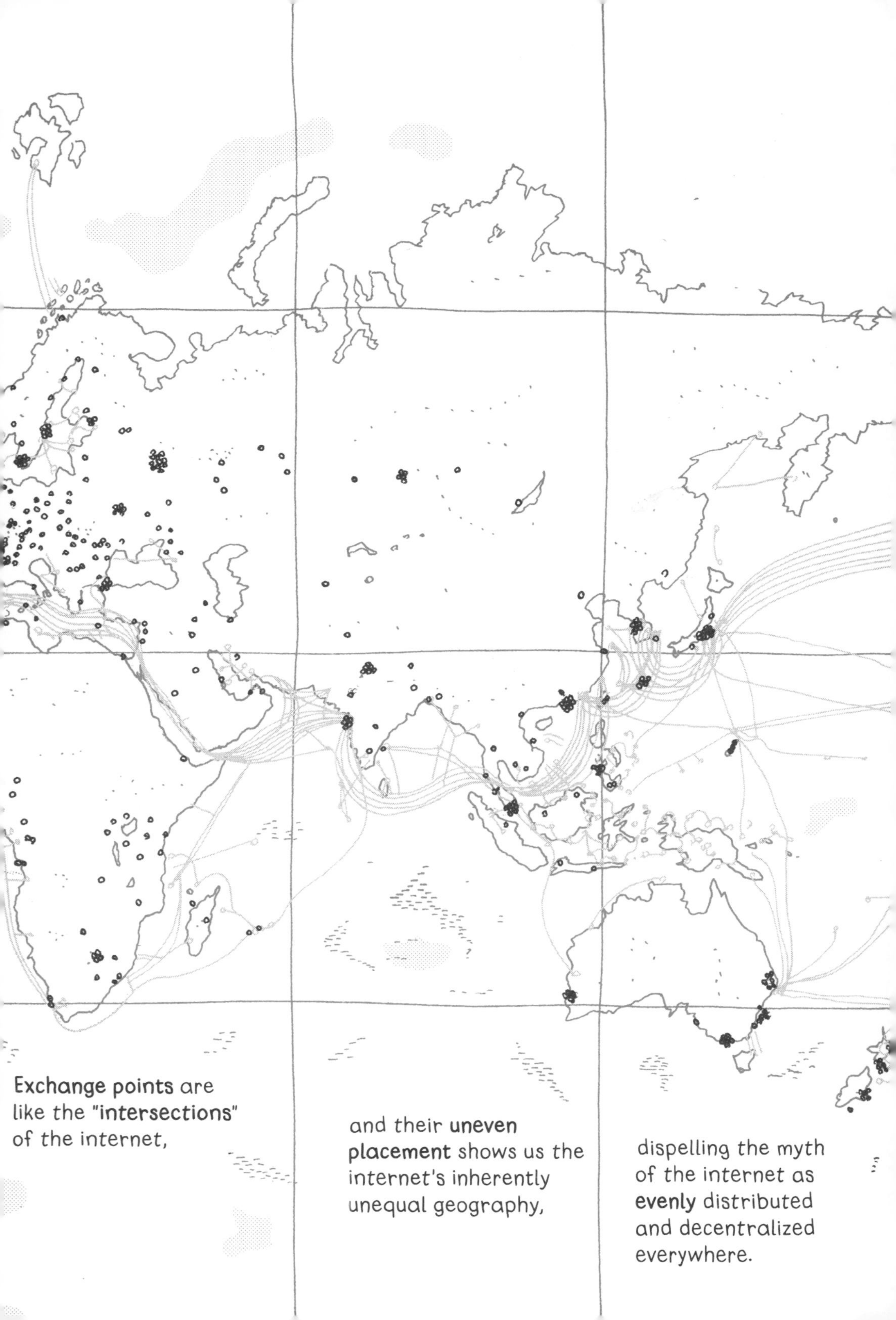

Exchange points are like the "**intersections**" of the internet,

and their **uneven placement** shows us the internet's inherently unequal geography,

dispelling the myth of the internet as **evenly** distributed and decentralized everywhere.

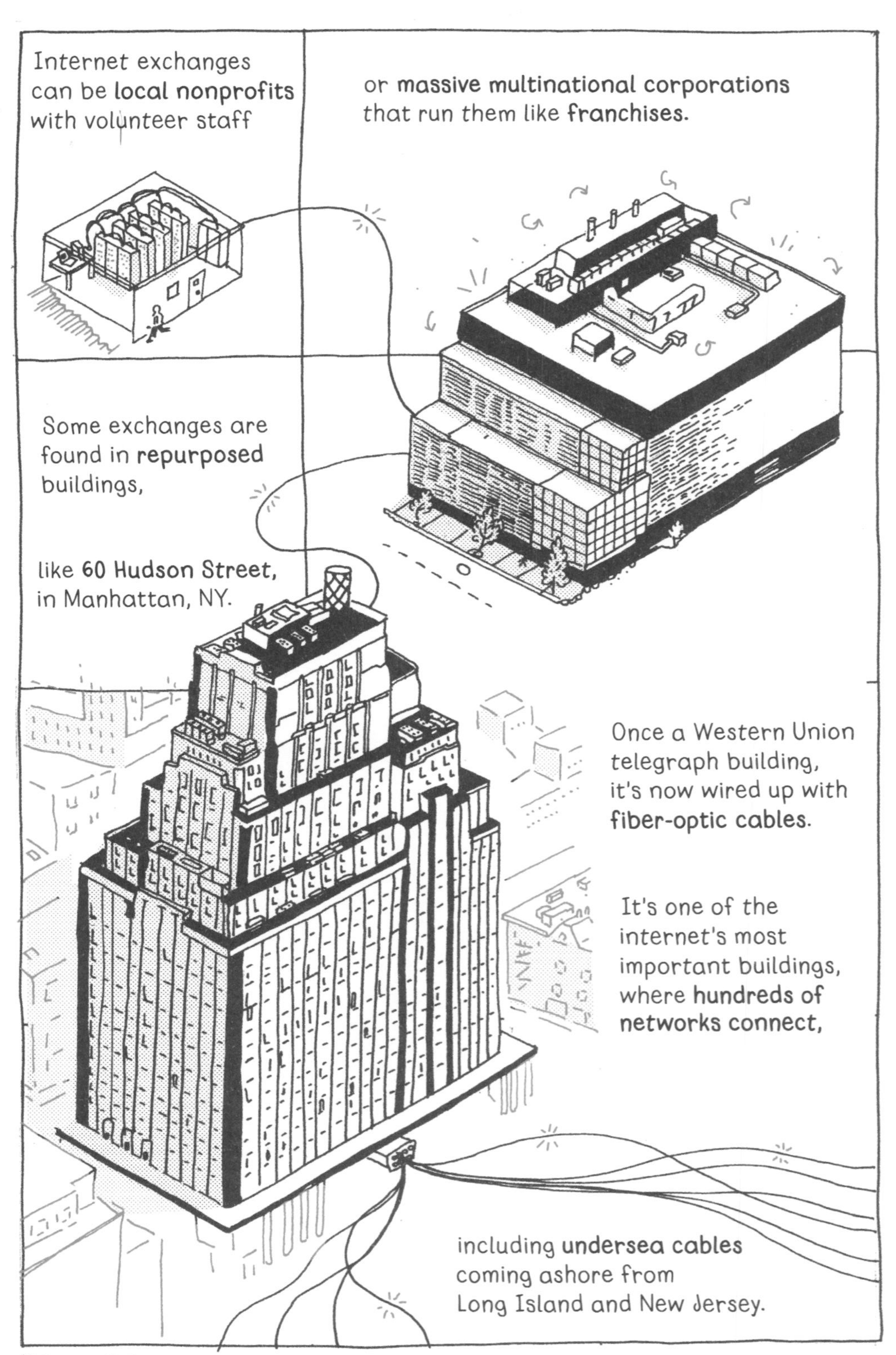
Internet exchanges can be **local nonprofits** with volunteer staff
or **massive multinational corporations** that run them like **franchises.**
Some exchanges are found in **repurposed** buildings,
like **60 Hudson Street,** in Manhattan, NY.
Once a Western Union telegraph building, it's now wired up with **fiber-optic cables.**
It's one of the internet's most important buildings, where **hundreds of networks connect,**
including **undersea cables** coming ashore from Long Island and New Jersey.

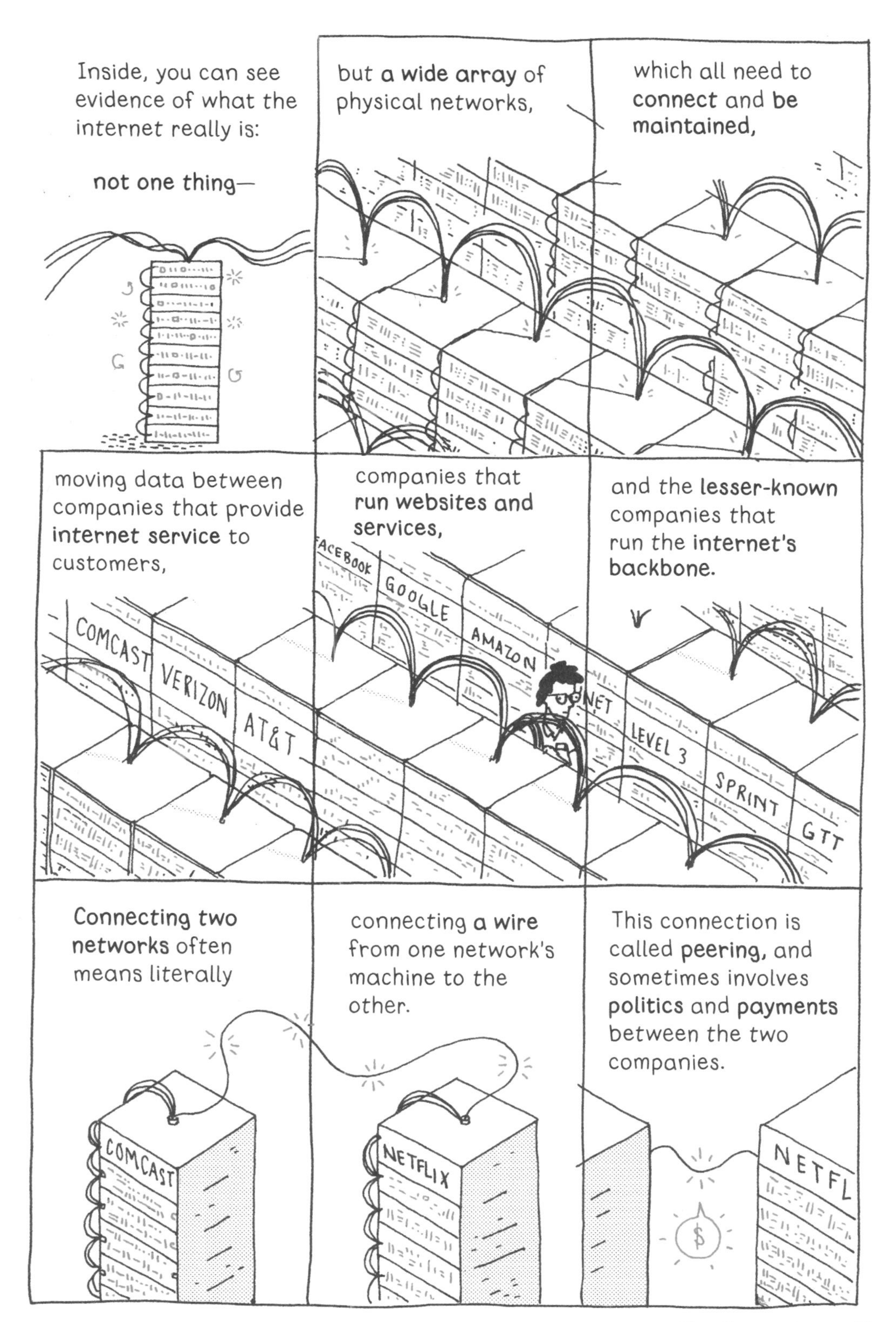

Inside, you can see evidence of what the internet really is:
not one thing—
but **a wide array** of physical networks,
which all need to **connect** and **be maintained,**
moving data between companies that provide **internet service** to customers,
COMCAST
VERIZON
AT&T
companies that **run websites and services,**
FACEBOOK
GOOGLE
AMAZON
and the **lesser-known** companies that run the **internet's backbone.**
NET
LEVEL 3
SPRINT
GTT
Connecting two networks often means literally
COMCAST
connecting **a wire** from one network's machine to the other.
NETFLIX
This connection is called **peering,** and sometimes involves **politics** and **payments** between the two companies.
NETFL
$

At the local level too, there are parts of the internet
Boston, MA
Verizon
where companies route the daily traffic
Pagosa Springs, CO
CenturyLink
of most towns and cities in the U.S.
Palmdale, CA
AT&T
at&t
These are small pieces of the internet
Maui, HI
Verizon, Hawaiian Telcom
coming in all shapes and sizes.
Jacksonville, GA
Windstream
in buildings often built for telephone switches
White River Junction, VT
Consolidated Communications
These secret buildings are ubiquitous,
Huntington, WV
Frontier
identifiable mostly by their uniquely low profile
Cheshire, CT
Frontier
and often a tiny network insignia.
verizon

From there, data either travels through underground ducts
or aerial cables
to reach modems and routers in people's homes.
Wireless data through 4G, LTE, or 5G
connects users to a nearby cell tower or antenna,
which is wired to the rest of the network through cables.
"Cellular" refers to the "cell"—the area served by an antenna.
Antennas are placed on the tops of buildings
and wherever else they can provide coverage.
When we post pictures, stream movies, or talk on the phone,
this is the path our data takes,
local building
modem + Router
regional exchange
but where does it come from?
And where does it go?

4.
Computers
"The cloud" became a popular concept in the mid-2000s,
but it has a long history.
References to the internet as a cloud are at least as old as the ARPANET,
when engineers drew amorphous, cloudlike shapes to abstract the geography of the networks.
NETWORK DIAGRAM 1977
ARPANET
But in practice, the principles behind cloud computing predate even the ARPANET.

The room-sized computers of the 1960s were a **huge expense,** and **inefficient** for just **one user.**

A process called **time-sharing** allowed multiple people to use a computer at once,

creating the illusion that the computer was their own.

As personal computers became affordable in the **1980s** and **'90s,**

the idea of having to **share** a computer became less relevant.

Everyone could run their **own programs** and store their data using their **own computer.**

Since the mid-2000s, how we use computers has changed **again,**

as we move much of what we do and data we own to the "**cloud.**"

Now the computers we use every day aren't the size of **rooms** . . .

BUZZ
BUZZ
BUZZ
Amazon data center,
one of dozens across
northern Virginia

Buzz
Buzz
Buzz
Buzz
They're the size of **warehouses.**

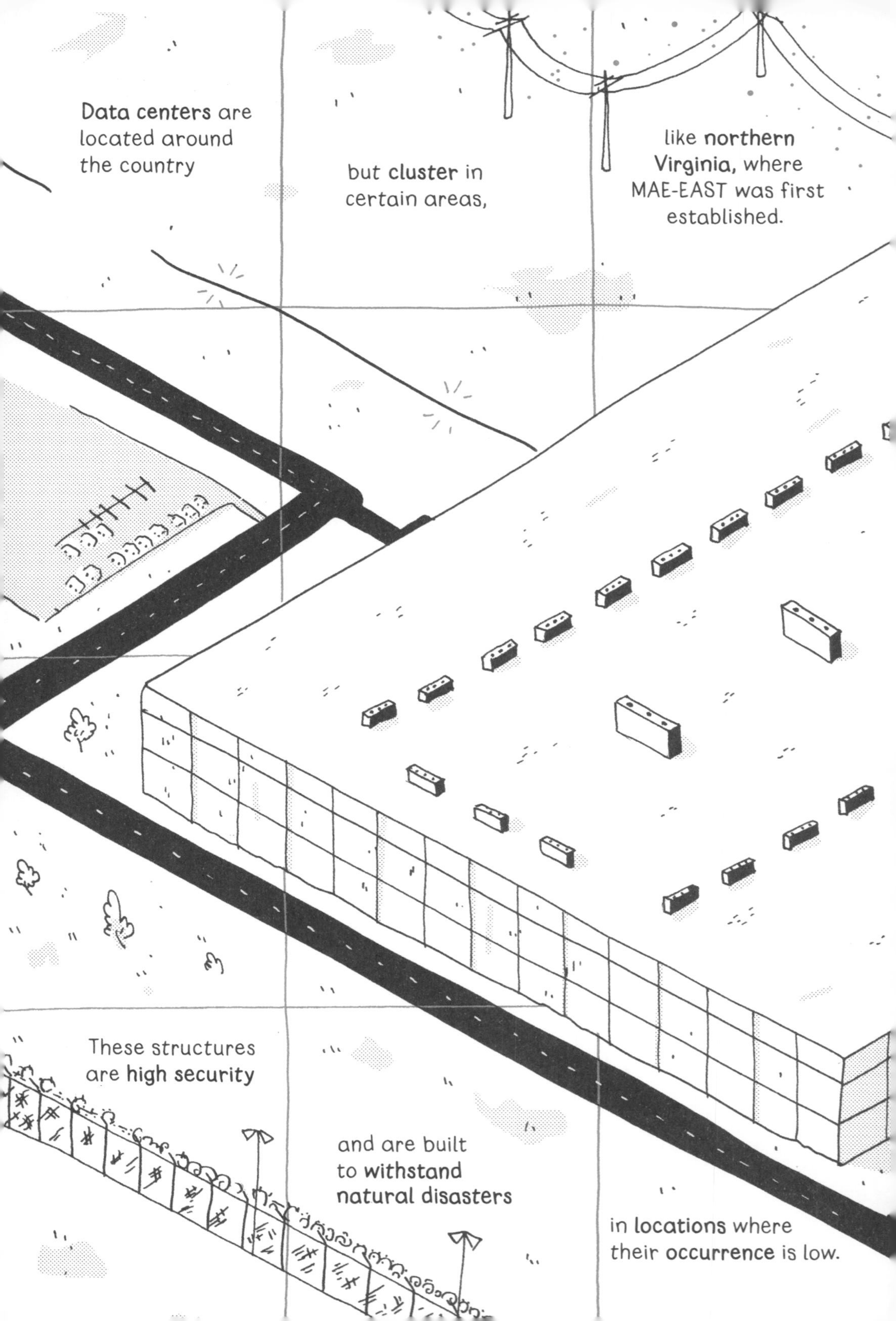
Data centers are located around the country
but **cluster** in certain areas,
like **northern Virginia,** where MAE-EAST was first established.
These structures are **high security**
and are built to **withstand natural disasters**
in **locations** where their **occurrence** is low.

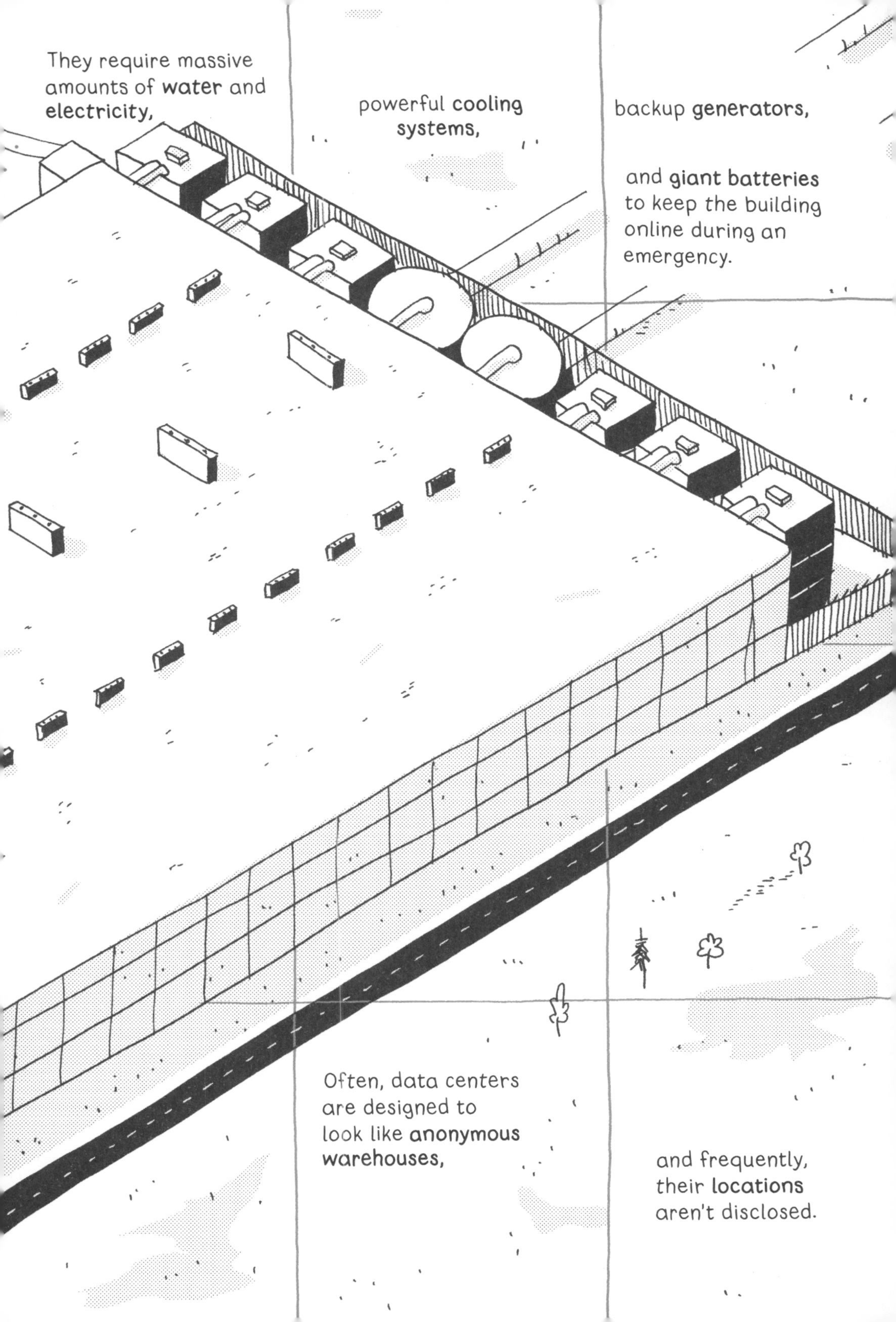
They require massive amounts of **water** and **electricity**,
powerful **cooling systems**,
backup **generators**,
and **giant batteries** to keep the building online during an emergency.
Often, data centers are designed to look like **anonymous warehouses**,
and frequently, their **locations** aren't disclosed.

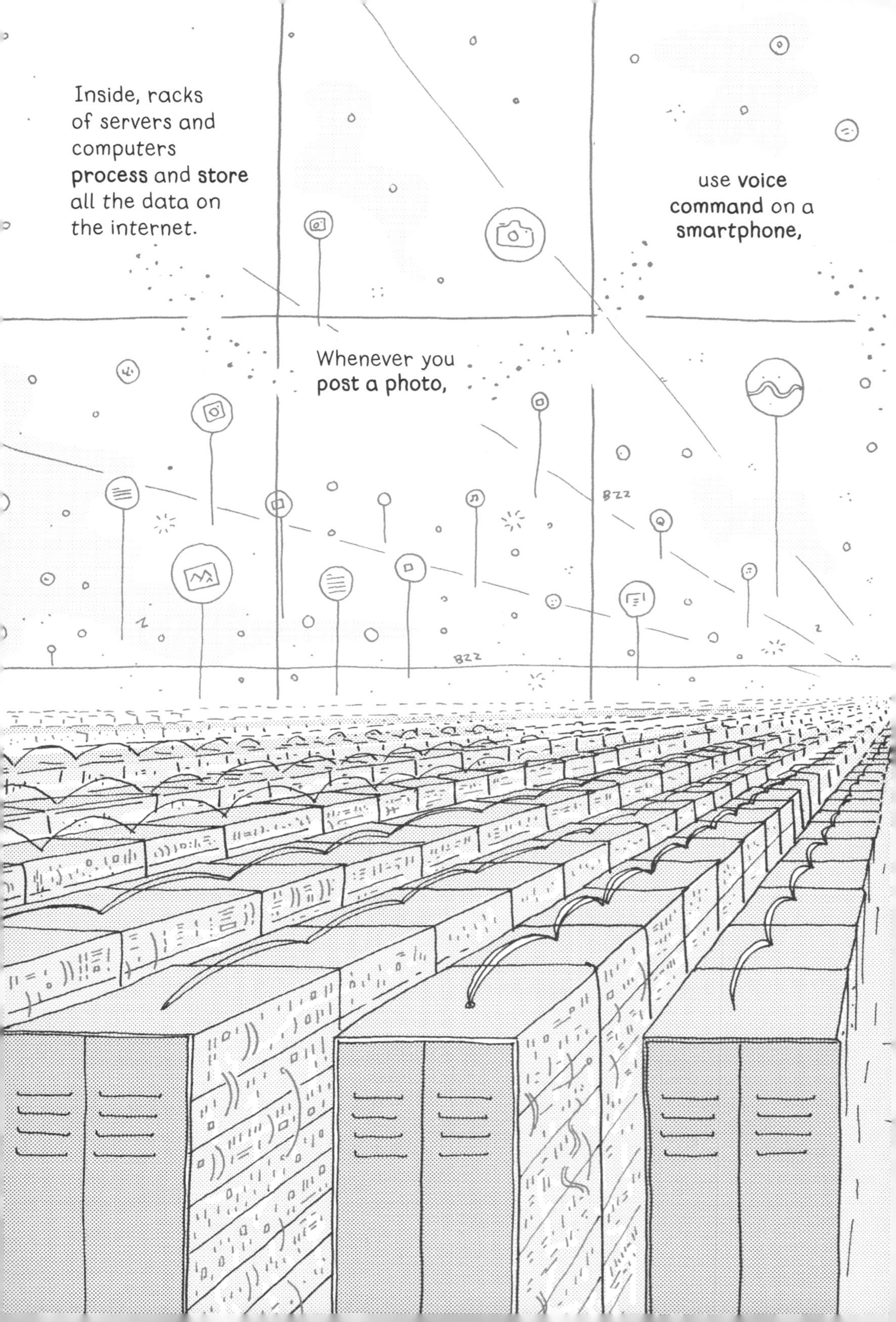
Inside, racks of servers and computers **process** and **store** all the data on the internet.
Whenever you **post a photo**,
use **voice command** on a **smartphone**,
BZZ
Z
BZZ
Z

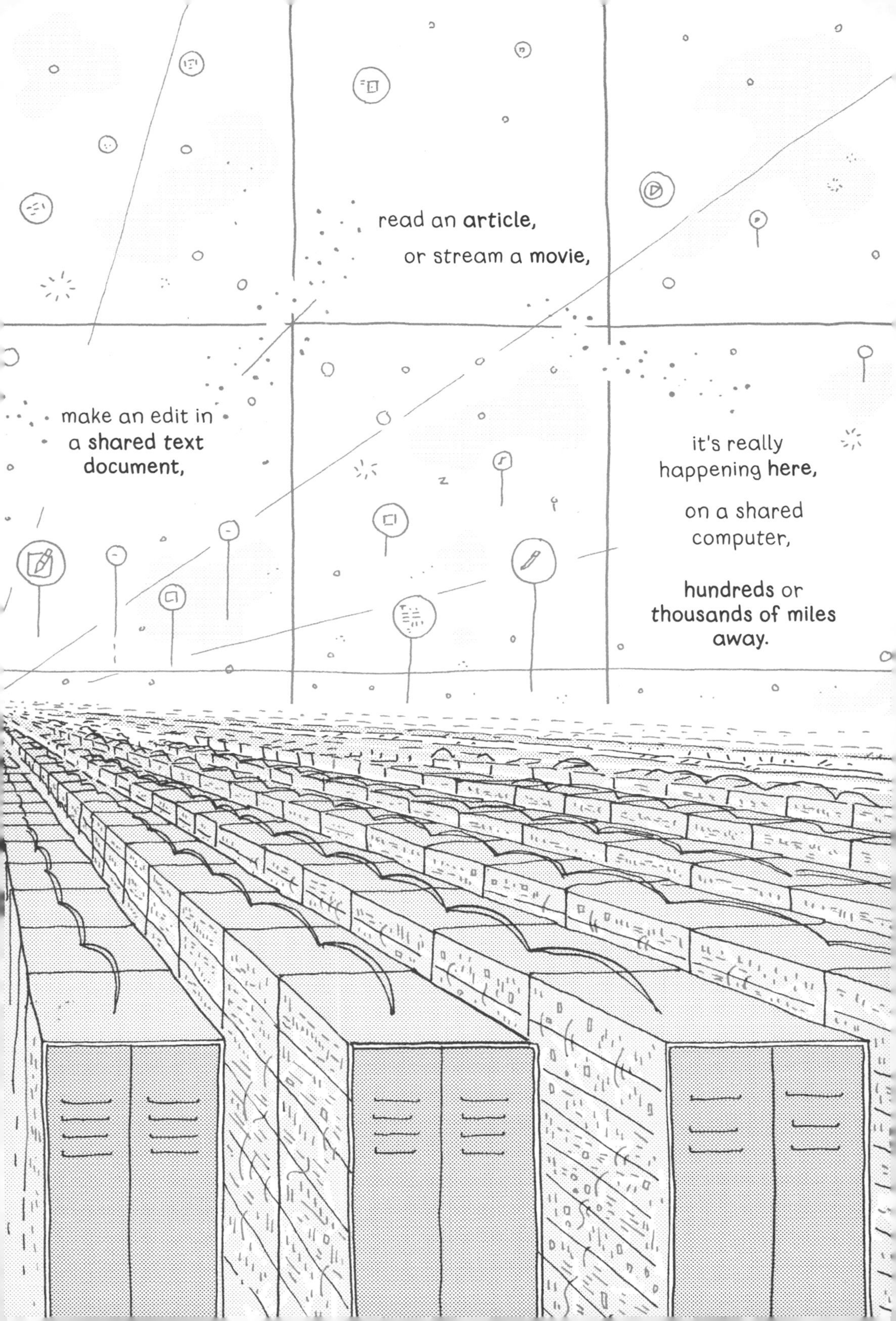

read an article,
or stream a movie,
make an edit in a shared text document,
it's really happening here,
on a shared computer,
hundreds or thousands of miles away.

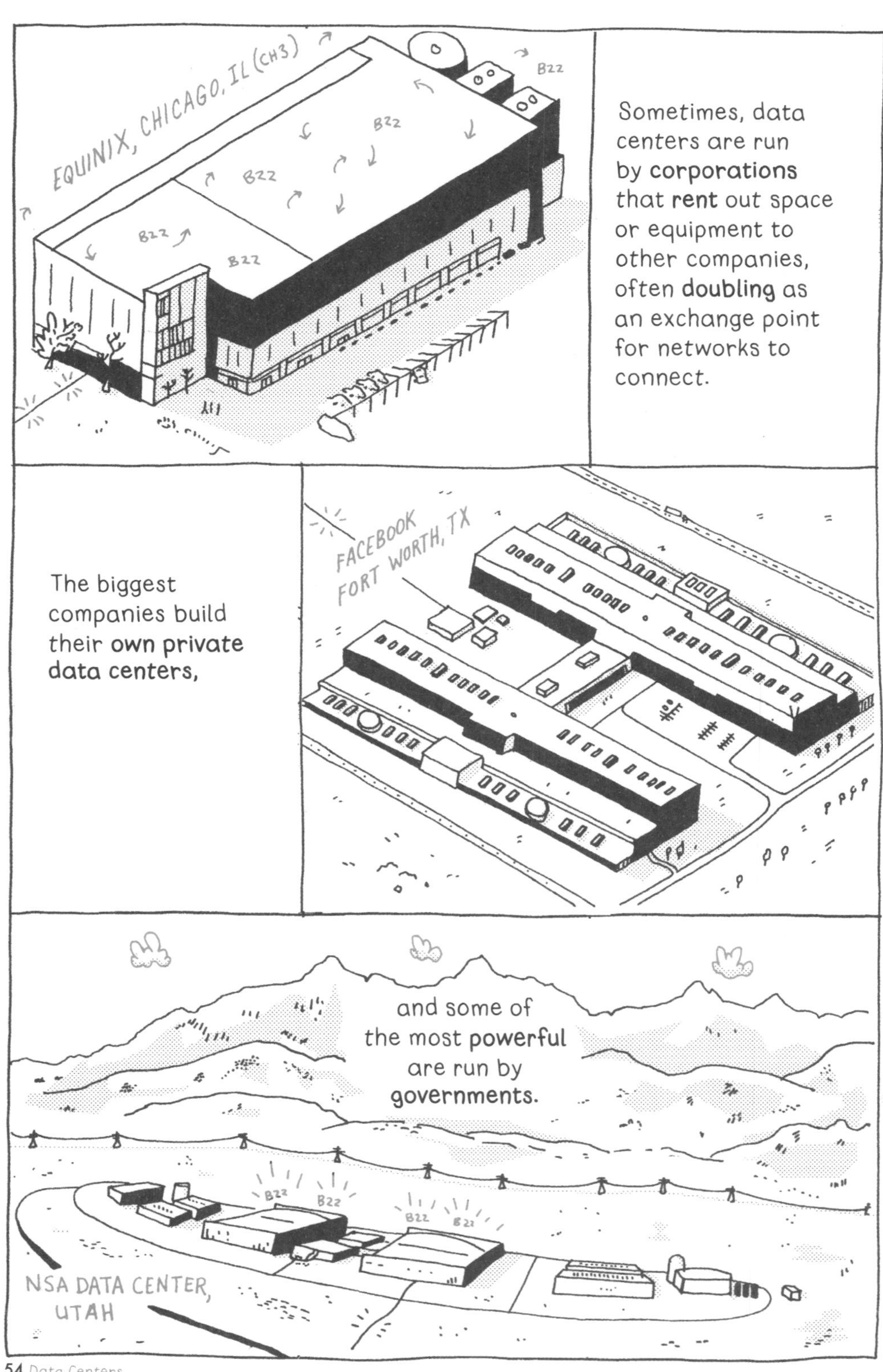
EQUINIX, CHICAGO, IL (CH3)
Sometimes, data centers are run by **corporations** that **rent** out space or equipment to other companies, often **doubling** as an exchange point for networks to connect.
The biggest companies build their **own private data centers,**
FACEBOOK
FORT WORTH, TX
and some of the most **powerful** are run by **governments.**
NSA DATA CENTER, UTAH

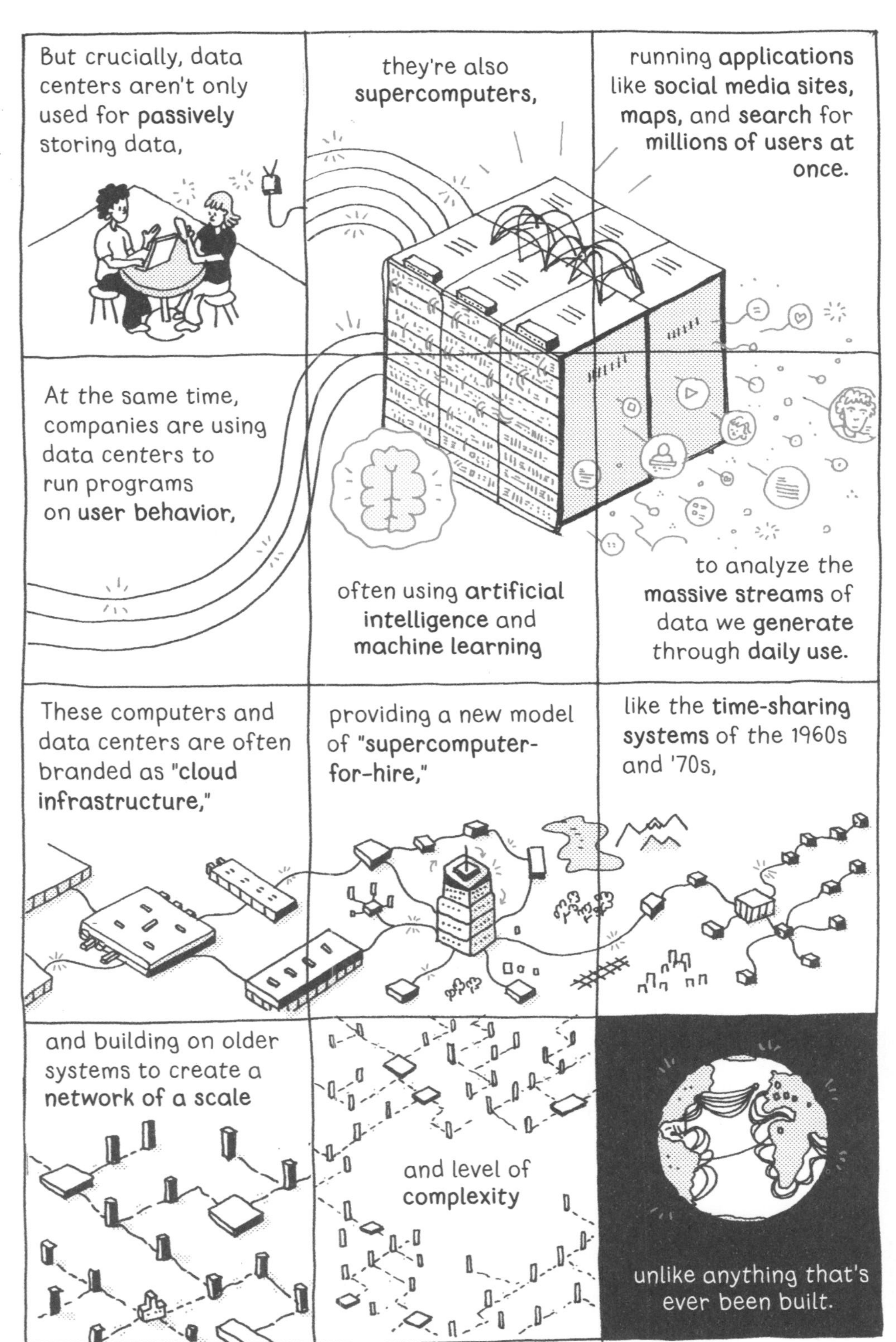
But crucially, data centers aren't only used for passively storing data,
they're also supercomputers,
running applications like social media sites, maps, and search for millions of users at once.
At the same time, companies are using data centers to run programs on user behavior,
often using artificial intelligence and machine learning
to analyze the massive streams of data we generate through daily use.
These computers and data centers are often branded as "cloud infrastructure,"
providing a new model of "supercomputer-for-hire,"
like the time-sharing systems of the 1960s and '70s,
and building on older systems to create a network of a scale
and level of complexity
unlike anything that's ever been built.

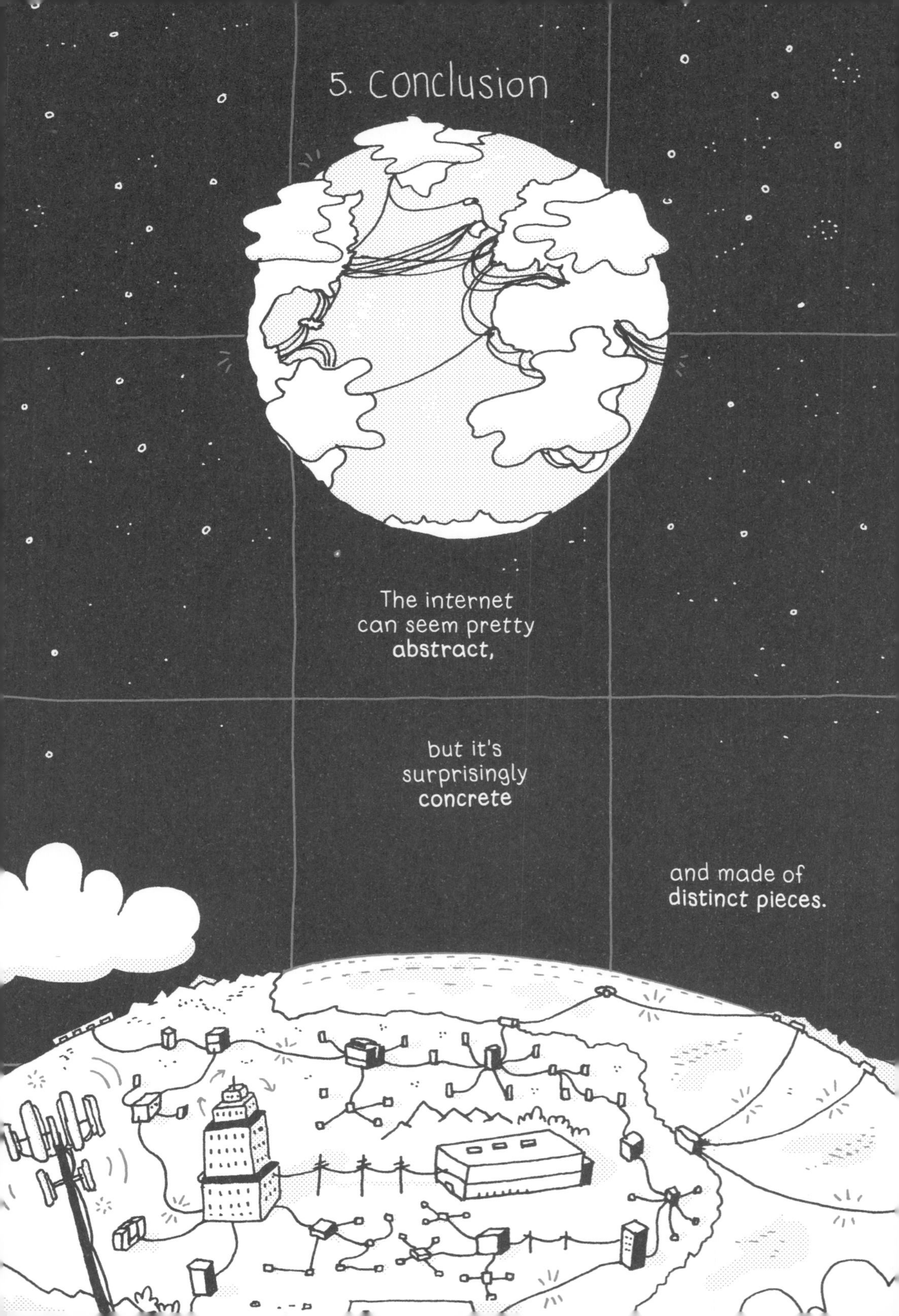

5. Conclusion
The internet can seem pretty **abstract,**
but it's surprisingly **concrete**
and made of **distinct pieces.**

Cables,

the "**tubes**" that transmit our communication—

at first carrying **one letter at a time** through copper telegraph cables,

now moving **terabytes of data** across oceans and continents **every second.**

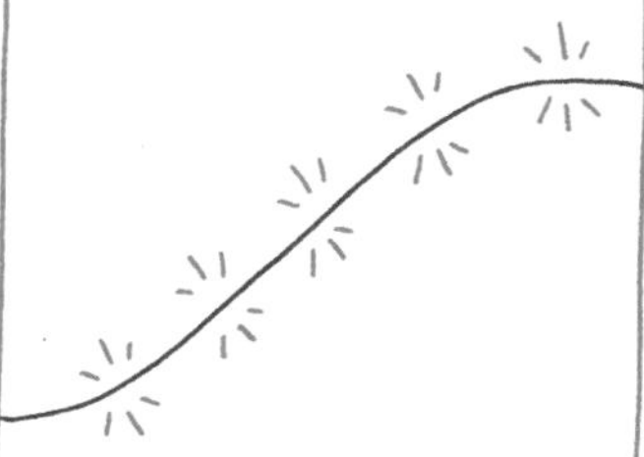

Exchanges,

the **intersections** that route our requests for information,

the internet's **clustered rotaries,**

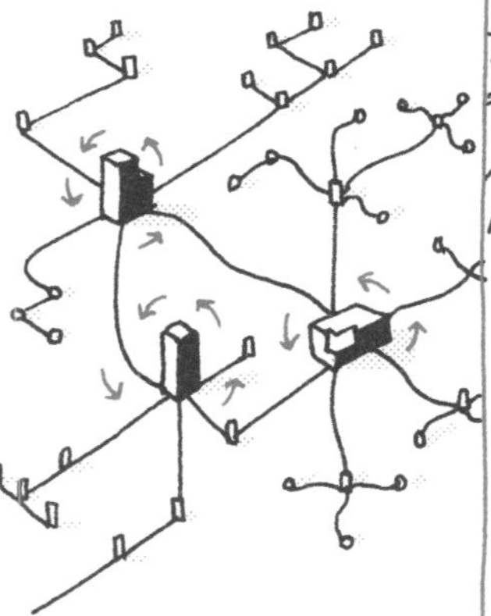

linking **thousands** of sources of information into the network.

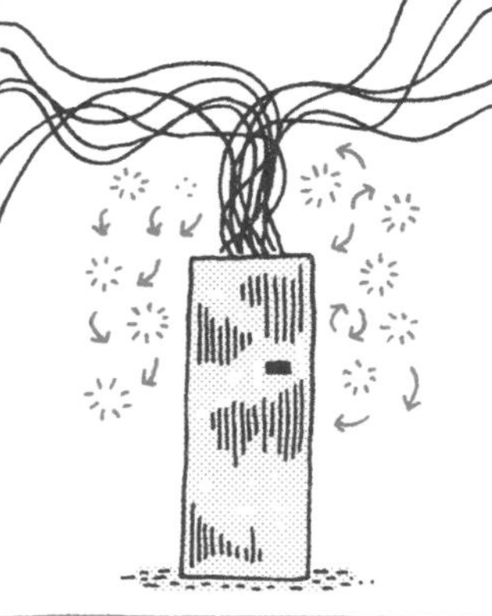

And data centers,

the "**clouds**" of the internet,

far-flung structures housing the shared computers used daily

by **billions** of people on **every continent.**

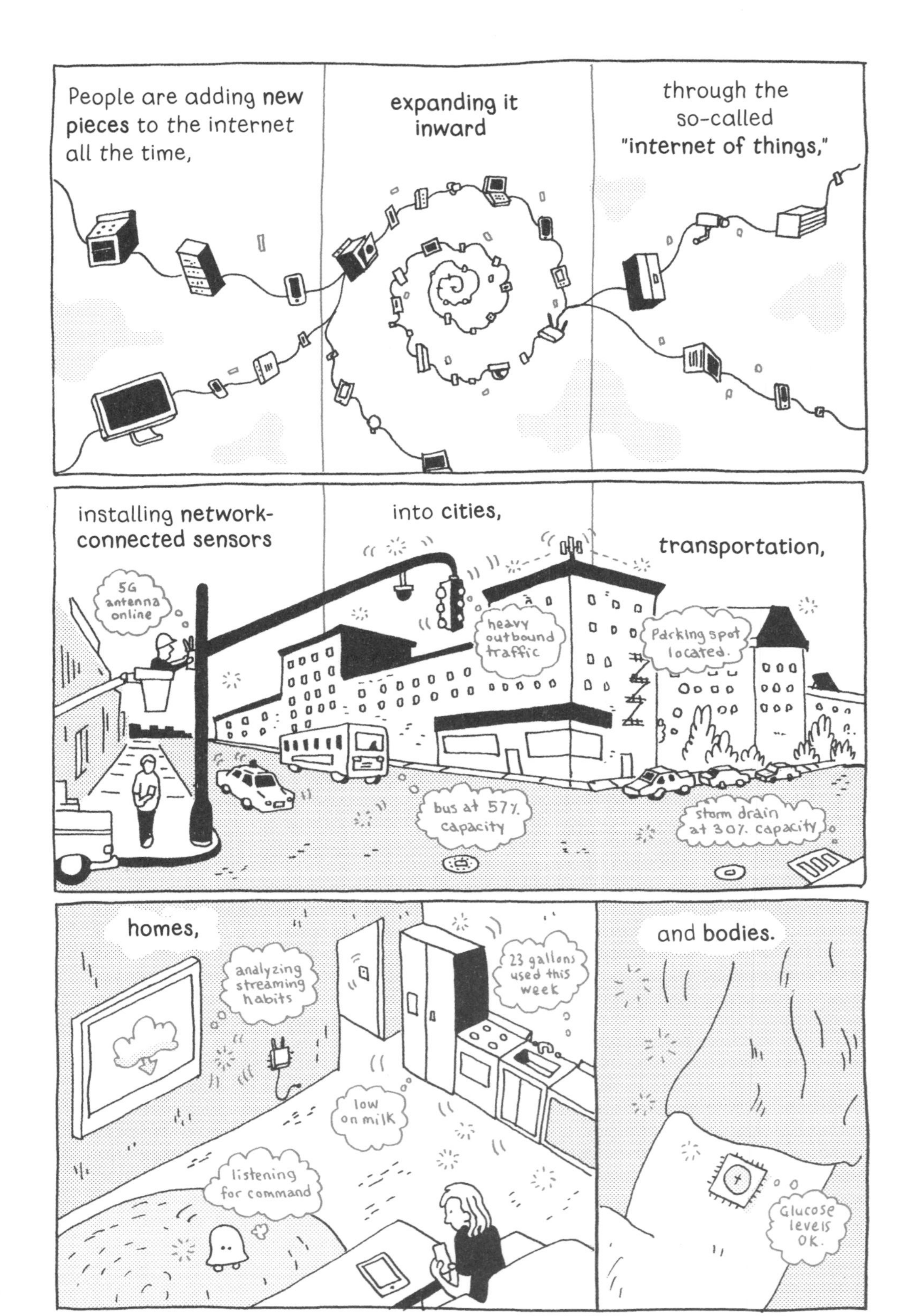
People are adding **new pieces** to the internet all the time,
expanding it inward
through the so-called **"internet of things,"**
installing **network-connected sensors**
5G antenna online
into **cities,**
heavy outbound traffic
transportation,
Parking spot located.
bus at 57% capacity
storm drain at 30% capacity
homes,
analyzing streaming habits
23 gallons used this week
low on milk
listening for command
and **bodies.**
Glucose levels OK.

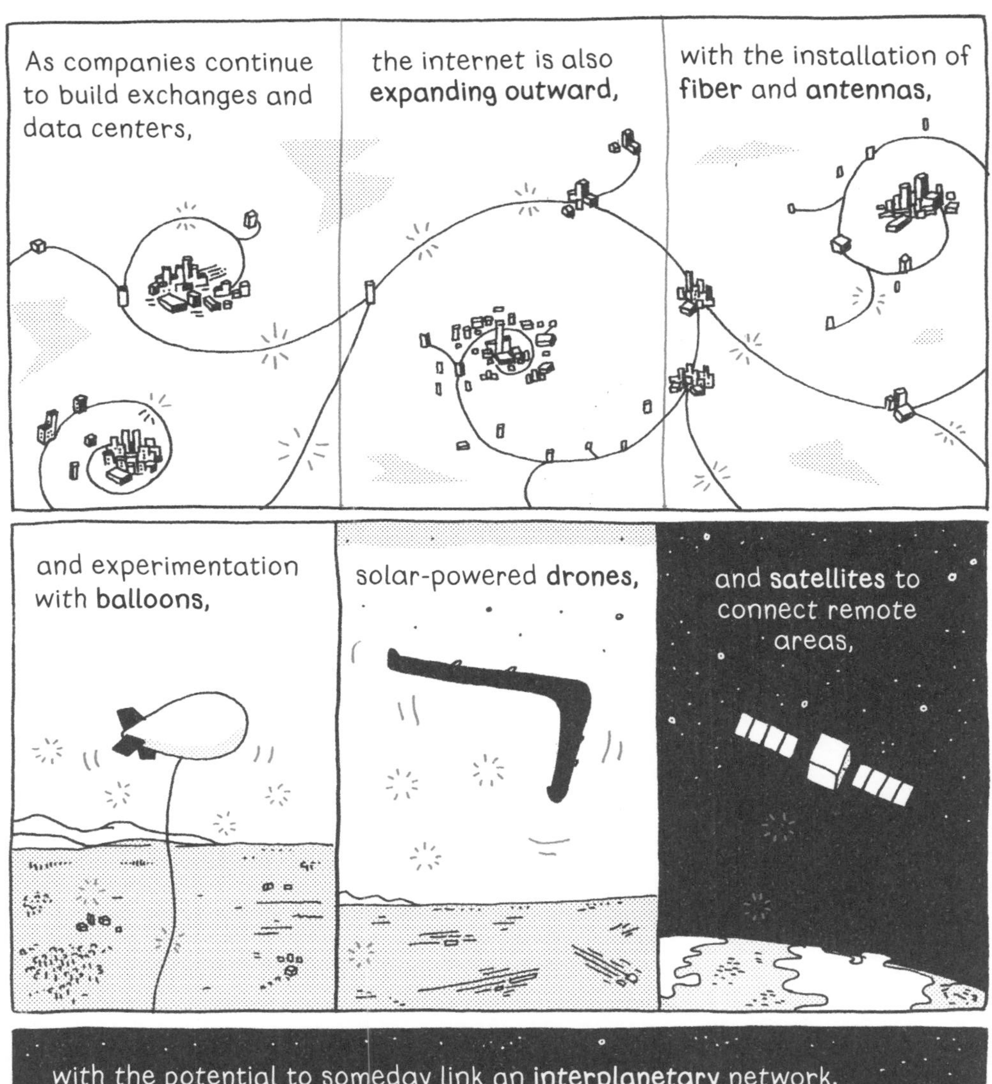
As companies continue to build exchanges and data centers,
the internet is also **expanding outward,**
with the installation of **fiber** and **antennas,**
and experimentation with **balloons,**
solar-powered **drones,**
and **satellites** to connect remote areas,

with the potential to someday link an **interplanetary** network.

Our metaphors tend to refer to **one piece** of the internet at a time.
But when you picture what all the parts look like **together,**
the internet looks a lot like a **computer,**
a rapidly evolving **neural network,**
wired into the surface of the Earth.

beep

The farthest Thunder that I heard
Was nearer than the Sky,
And rumbles still, though torrid Noons
Have lain their Missiles by –
The Lightning that preceded it
Struck no one but myself –
But I would not exchange the Bolt
For all the rest of Life –
Indebtedness to Oxygen
The happy may repay,
But not the obligation
To Electricity –
It founds the Homes and decks the Days
And every clamor bright
Is but the gleam concomitant
Of that waylaying Light –
The Thought is quiet as a Flake –
A Crash without a Sound,
How Life's reverberation
Its Explanation found –

—Emily Dickinson

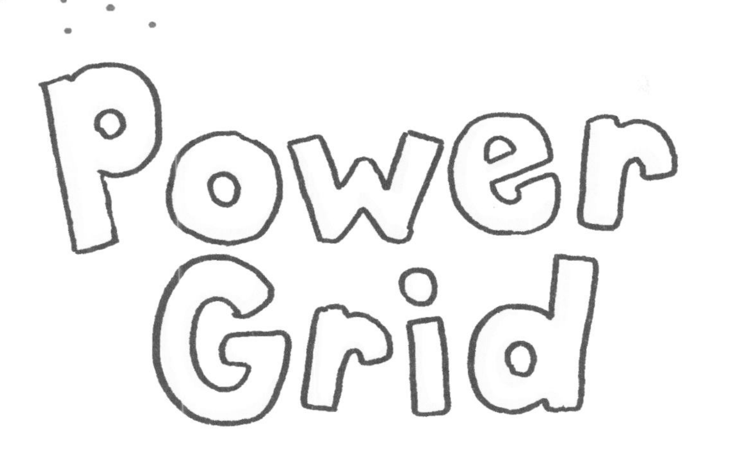

Power Grid

How do we power our world
with electricity?

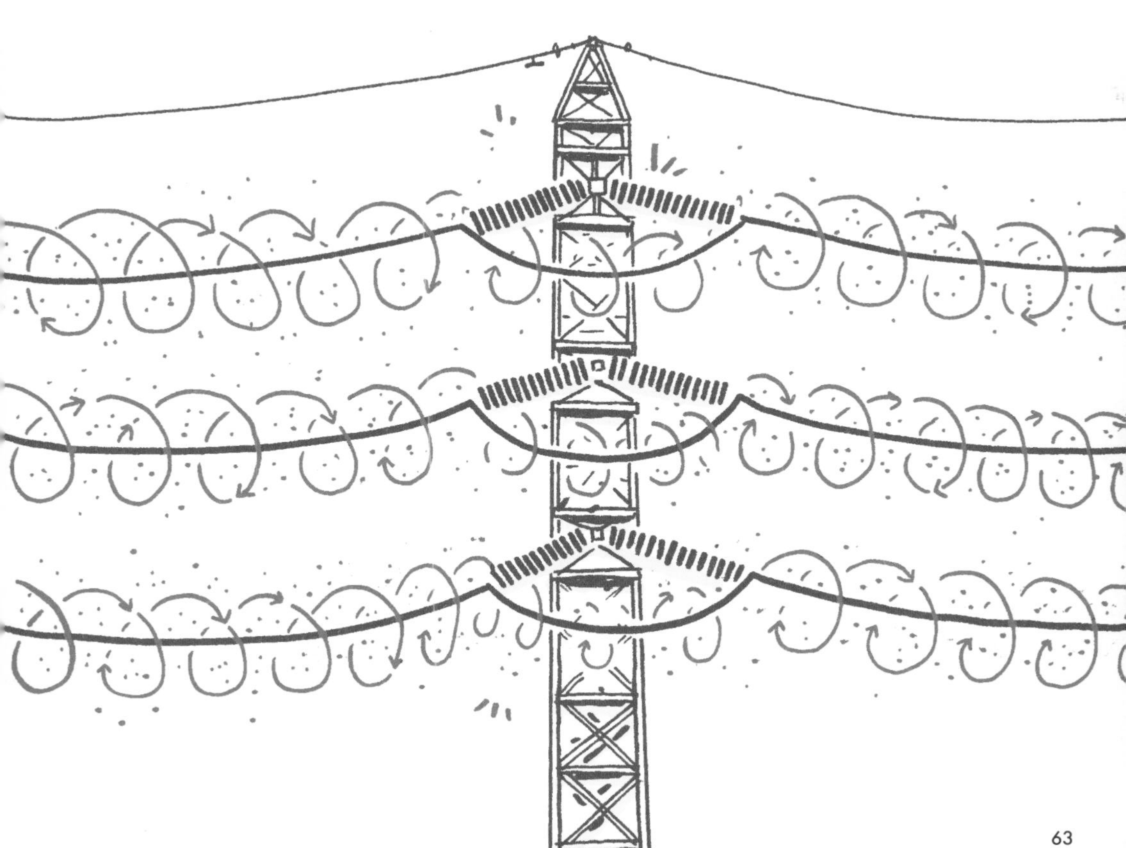

CLICK
GAS
5G

Electricity is something
we don't really think about

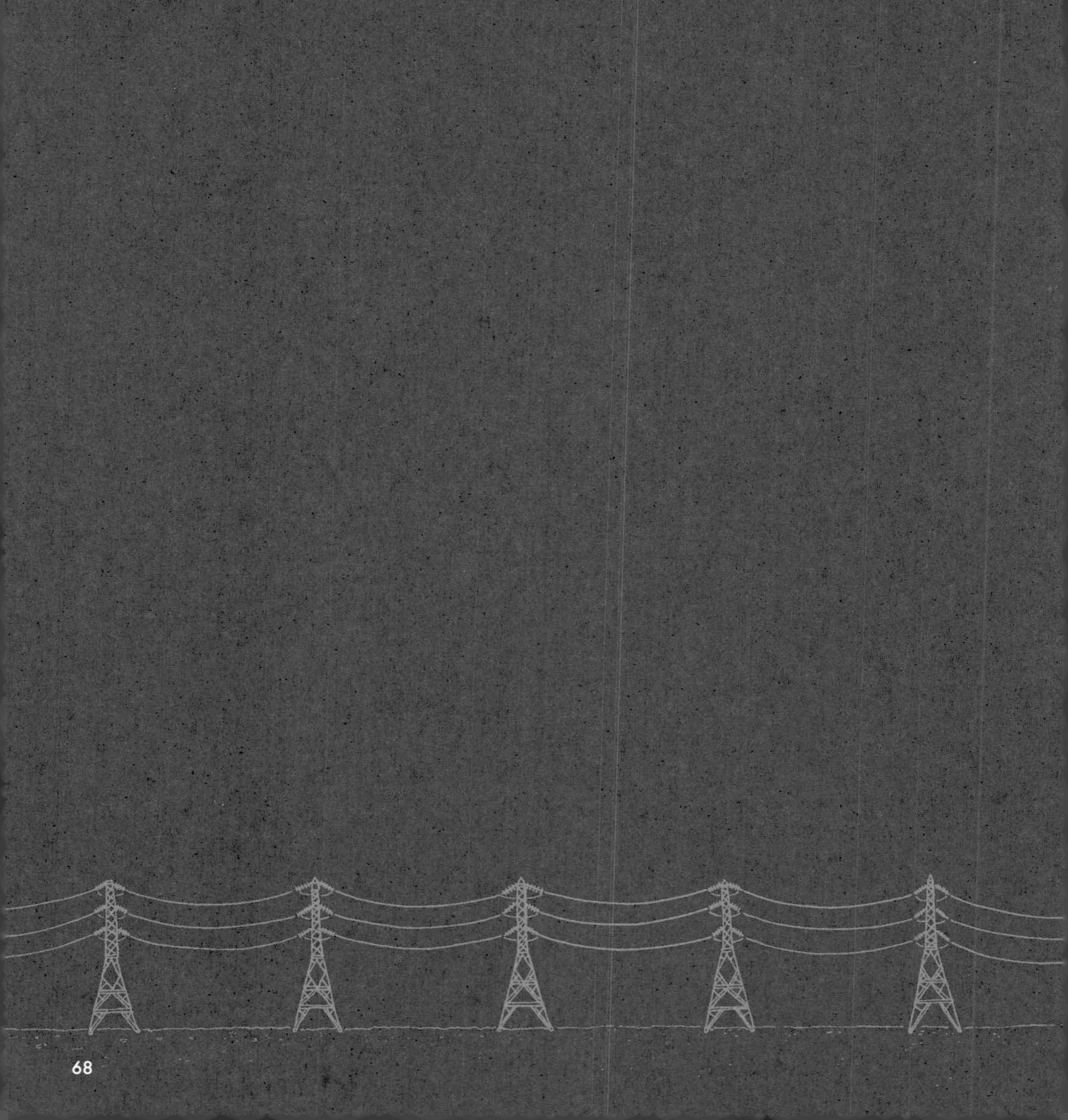

until we don't have it.

During a **blackout,**
we realize how much of our daily lives **rely** on electricity.
Wow, uh, each thing that I own has suddenly become useless garbage.
And more importantly, how essential electric power is
for our society to run **smoothly**
and **safely** for everyone.
HONK
HONK
HONK
HONK
HONK
HONK
GAS
CLOSED

Even in our **brief moments** without it,
My phone died.
RIP Phone
You need a phone with more battery life.
we talk about **electricity** like it's a **life force**—
that can **animate** everything
from **factories**
to **microchips.**
Electricity is our **constant companion.**
is there an outlet here?
over there.
But for something so central to daily life for most of us,
how often do we think about what it actually is?
hmm...

It may seem basic, but "**What is electricity?**" is not an easy question.
Energy? Is that the same thing?
(sort of— electricity is a type of energy.)
It comes from the sky?
Electricity is hard to **picture** because, aside from lightning,
it's mostly **invisible**.
(and it's usually represented with symbols.)
And it's hard to **describe** because there's nothing to **compare** it to.
Sooo... magic?
(basically...)

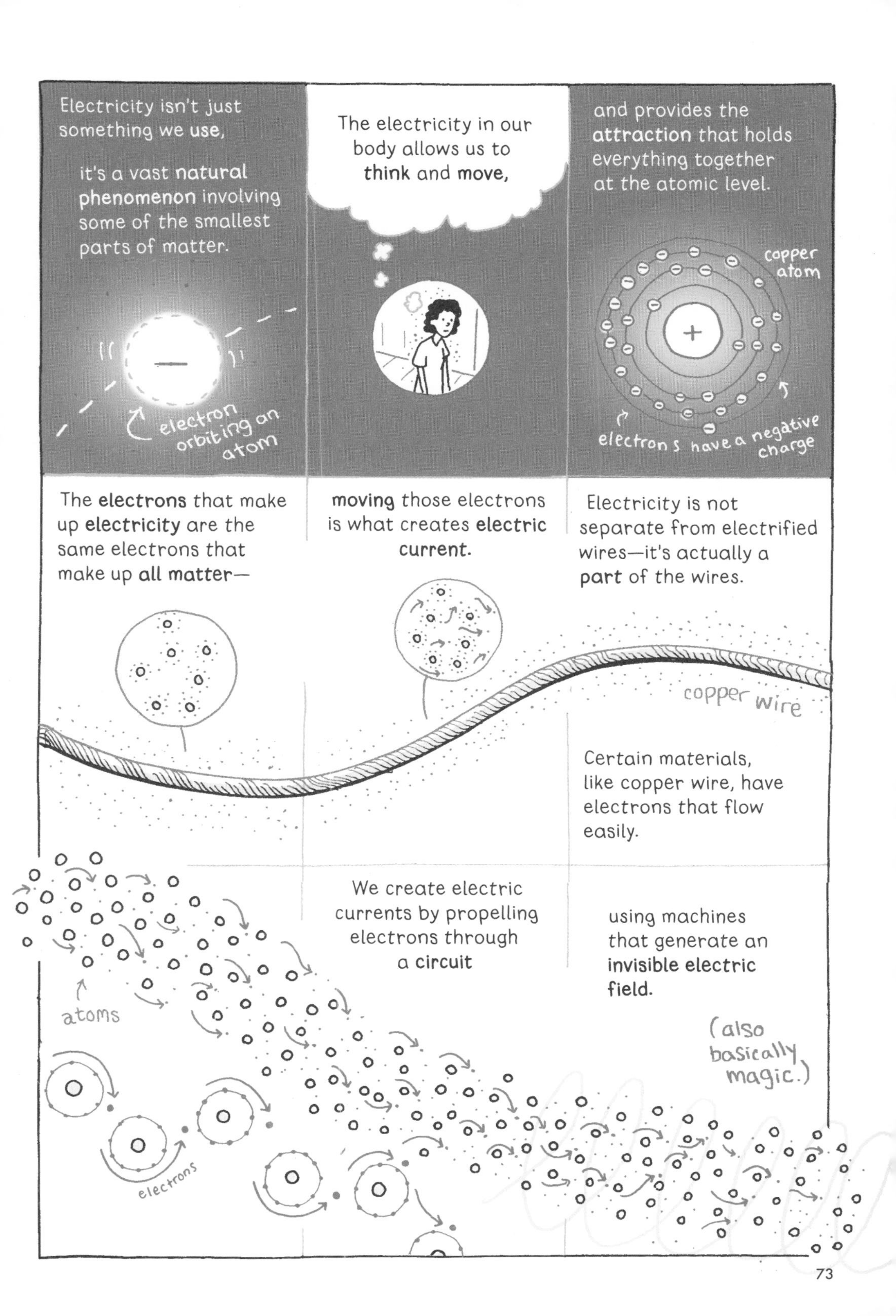
Electricity isn't just something we **use**,
it's a vast **natural phenomenon** involving some of the smallest parts of matter.
electron orbiting an atom
The electricity in our body allows us to **think** and **move**,
and provides the **attraction** that holds everything together at the atomic level.
copper atom
electrons have a negative charge
The **electrons** that make up **electricity** are the same electrons that make up **all matter**—
moving those electrons is what creates **electric current.**
Electricity is not separate from electrified wires—it's actually a **part** of the wires.
copper wire
Certain materials, like copper wire, have electrons that flow easily.
atoms
electrons
We create electric currents by propelling electrons through a **circuit**
using machines that generate an **invisible electric field.**
(also basically magic.)

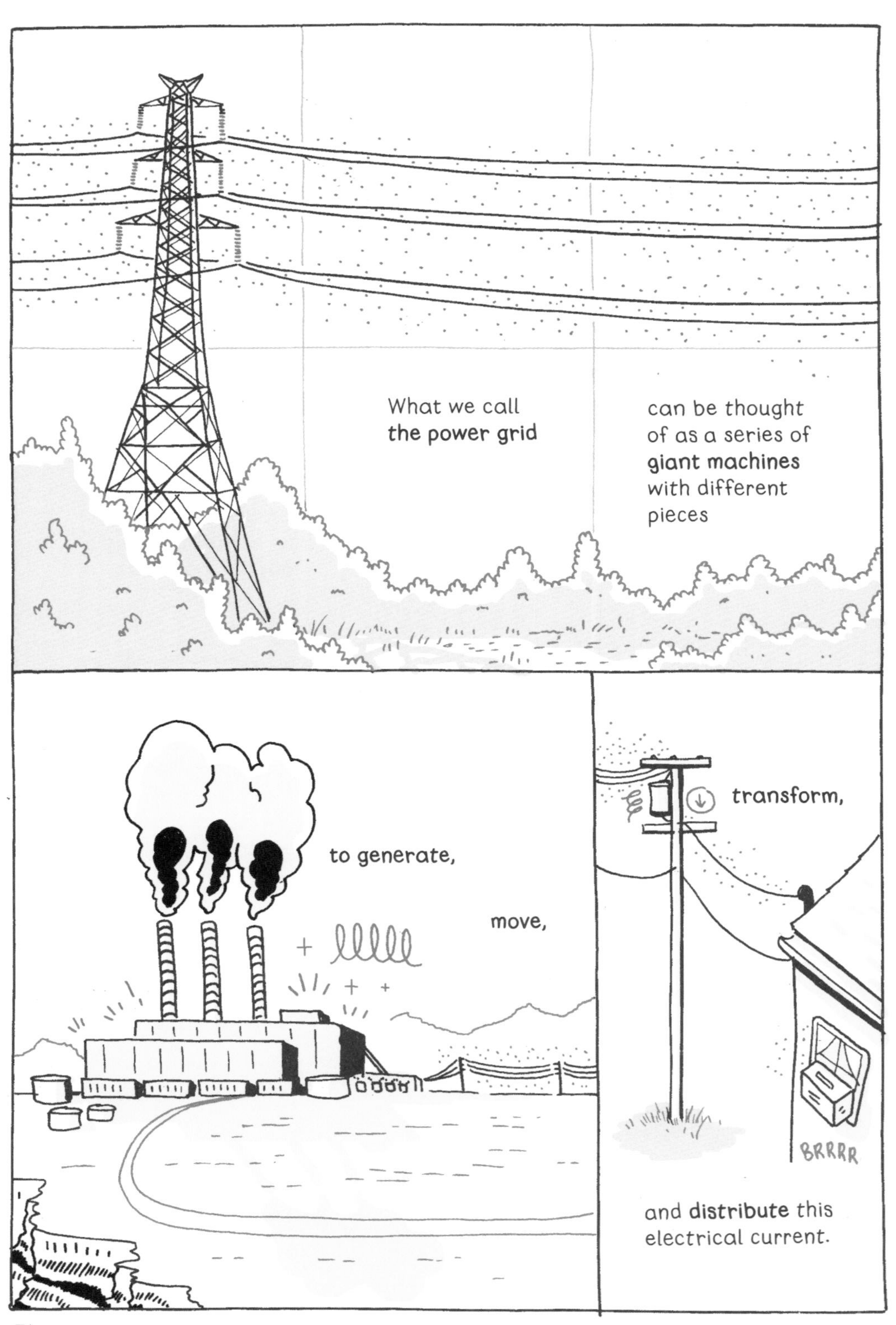
What we call **the power grid**
can be thought of as a series of **giant machines** with different pieces
to generate,
move,
transform,
BRRRR
and **distribute** this electrical current.

While we take it for granted today,
this system that powers so much of our world
didn't exist a few generations ago.
This machine had to be **imagined,**
invented,
and installed,
piece by **piece,**
over more than **100 years.**
1890
1900
1910
1920
1930
1940
1950
1960
1970
1980
1990
2000
2010
2020
2030

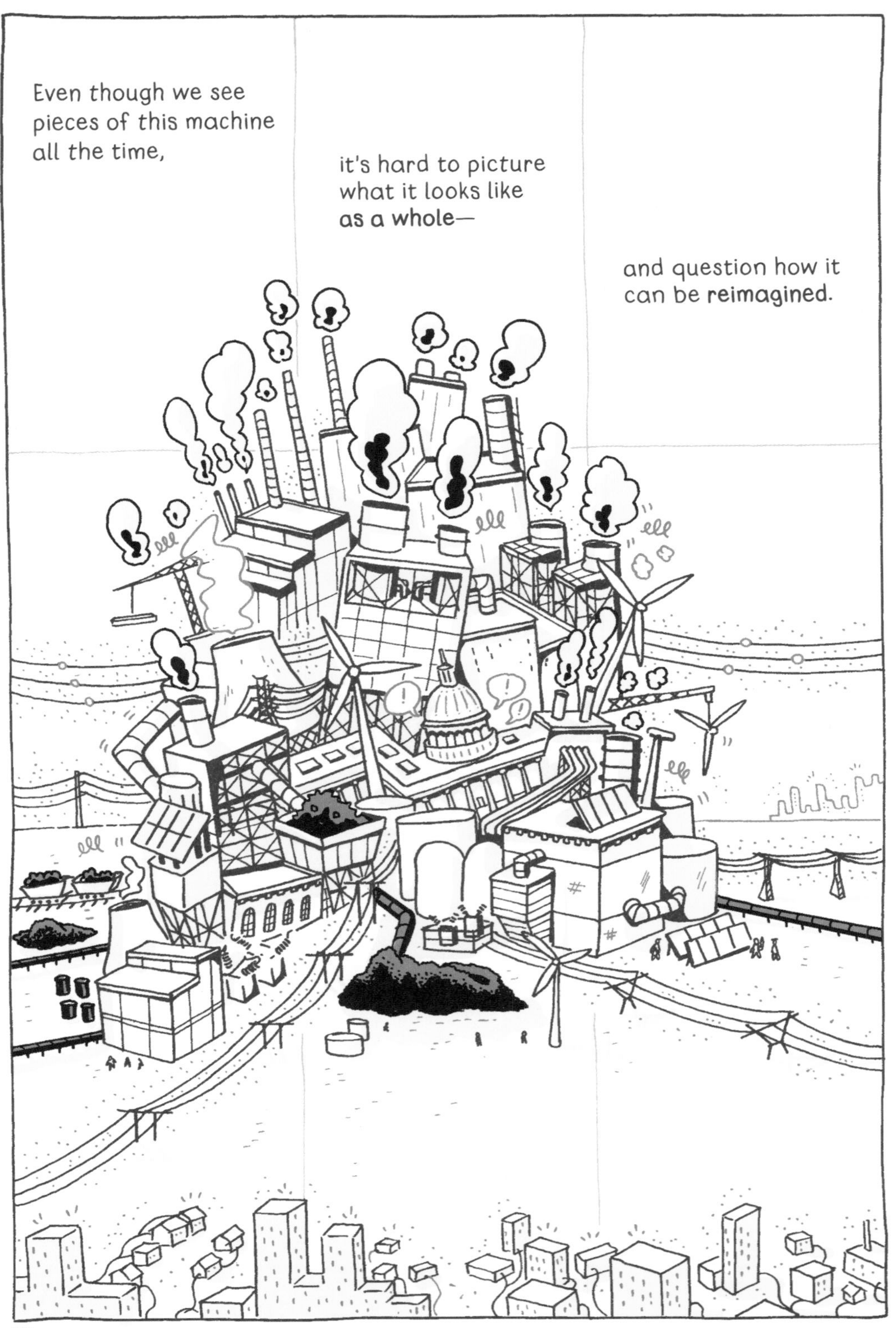
Even though we see pieces of this machine all the time,
it's hard to picture what it looks like **as a whole**—
and question how it can be **reimagined**.

After **100 years,** the way we **make** and **use** energy needs to change.
Our systems are creating catastrophic changes in the climate,
threatening entire populations and upending the lives of the planet's most vulnerable people.
In an era of climate change,
we talk about using **"100% renewable energy"** and **"greening our grid,"**
but we talk less about the system that does exist, and what these actions require.
STOP BURNING OUR FUTURE
Clean Energy Revolution
BAN FRACKING
Future Power
NO MORE COAL
CLIMATE LEGISLATION NOW
There's No Planet B
In order to really reimagine the grid,
it's helpful to understand the system we do have
and how we got it.
?
???

Experiments and Invention

For most of our history,
humans had **no idea** what electricity was.
Rumble
Rumble
Rumble
ZAP
But we've always harnessed **energy** in a variety of ways.
Someone should look into that someday.
EXPLODE
Energy is transferred all the time.
For example, plants absorb the sun's energy to grow,
nom
nom
which is transferred when it's consumed.
nom
nom
nom
And **so on**,
gulp
and **so on** . . .
nom
nom
nom
nom

Long before electricity, we used these **energy transfers** to do **work for us.**
like the energy in moving water
to turn gears,
energy in moving air
to grind grain
energy in the muscles of animals
to pull & carry weight
Life could be better.
And we've used a range of fuels for **heat and light.**
energy stored in plants
ooh, game changer.
energy stored in oil
burns very dirty
cough cough
I do *think* that this sperm whale fuel burns cleaner and brighter than all the rest!
So long, whales.
energy stored in whale head
When curious "**natural philosophers**" began looking into strange sparks and unexplained attractions,
what are these odd electrical virtues?
Otto Von Guericke, 1672
they had little idea what the effects **meant,**
or that it might be a **source of energy.**

Our first practical machine for electricity was called a Leyden jar.
Europe, 1745
Basically, it was a jug that could store static electricity and zap things,
and at the time, it was somewhere between cutting-edge science and magic.
!
Whoaaaa...
Over the next century, scientists built on (and took credit for) each other's electrical discoveries.
Alessandro Volta created a charge by stacking zinc and copper discs—
inventing the first battery.
chemical energy
to electric energy
And Michael Faraday created an electric current by rotating a conductor past magnets—
the first generator.
(mechanical energy to electric)
Soon, practical uses for electricity emerged,
like the telegraph,
tap tap tap
Beep Beep Beep
1840s
and arc lights, which used exposed electricity to create a blinding light,
"moon-tower"
(mostly for outdoors)
this light makes me look awful.
1880s
best used for lighting streets or large areas.
But no one could figure out a way to make a practical, softer electric lamp for inside a home.

By the late 1870s, Thomas Edison was one of many trying to make a light bulb that lasted longer than 30 seconds before burning out.
Edison was a young upstart, and after the success of his phonograph,
he aimed to make a "minor invention every ten days," and a "big thing every six months or so."
dropped out of school
mostly deaf
POP!
first commercial research lab in the U.S.
Using the expertise of others, and a lot of trial and error,
he would stay up all night inventing with a team.
They tried thousands of designs and materials for the light bulb,
ordering plants from across the world to find something that would glow when electrified,
eventually settling on carbonized cotton.
creates soft orange light
But the bulb was just the start.
Edison knew there needed to be a whole system
for generating, transmitting, and metering the electricity required for the lights.
He and his team invented more than just the light bulb . . .

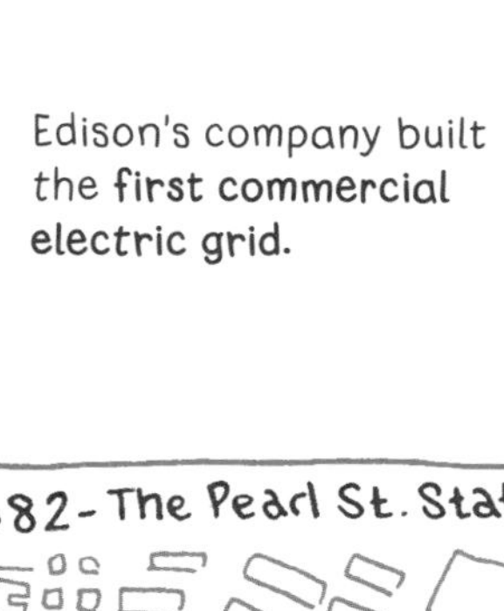

Edison's company built the **first commercial electric grid.**

Using one **central station**, it lit thousands of bulbs across a **one-mile area** in lower Manhattan,

strategically serving prominent **banks and newspapers** like the *New York Times.*

They also had to figure out how to **transport electricity** in the city using **parallel circuits.**

Edison's workers dug up **fourteen miles** of city streets and buried **80,000 feet** of copper cable **underground,**

due to widespread **fears about the safety** of this new energy.

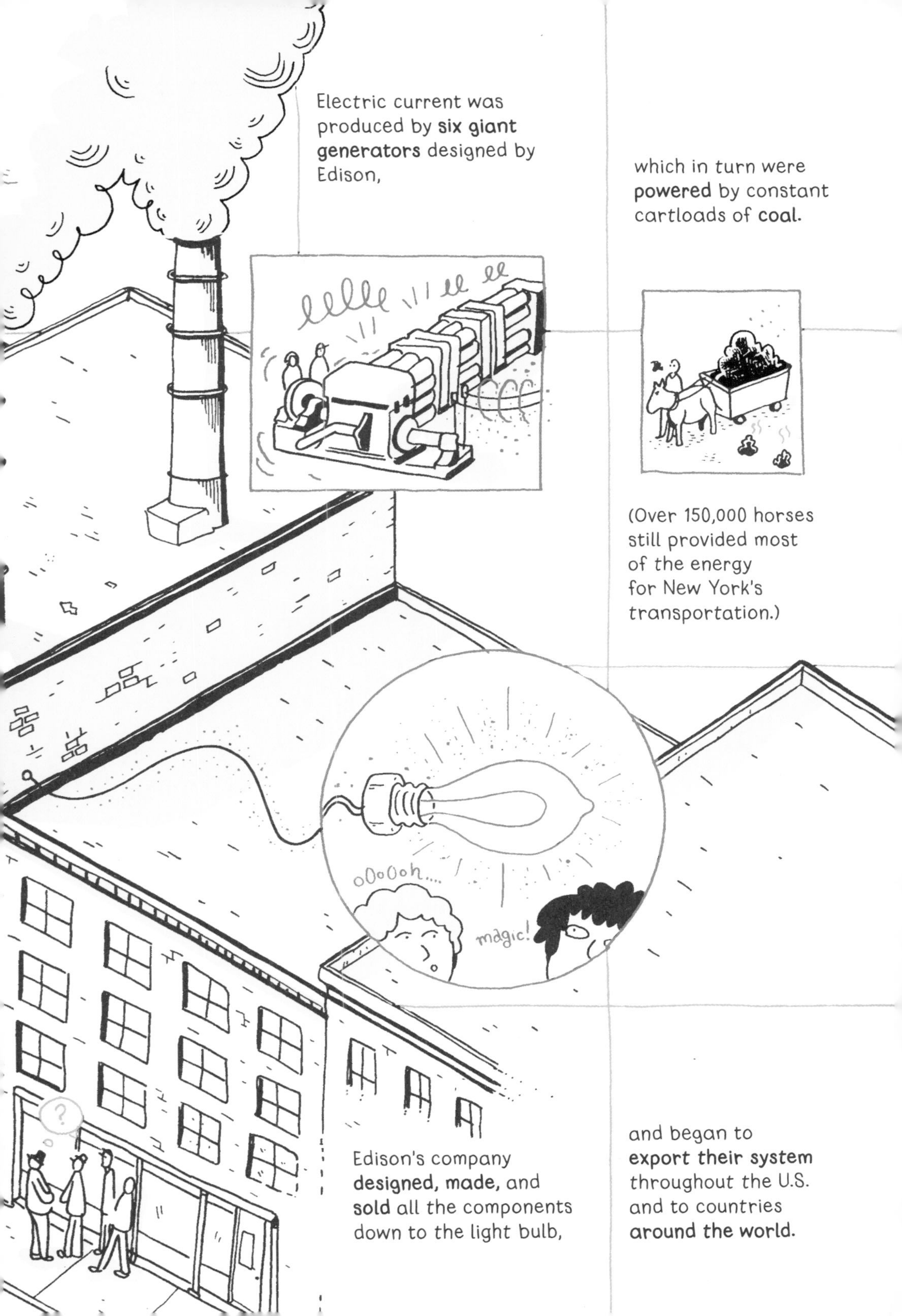
Electric current was produced by **six giant generators** designed by Edison,
which in turn were **powered** by constant cartloads of **coal.**
(Over 150,000 horses still provided most of the energy for New York's transportation.)
oOoOoh....
magic!
?
Edison's company **designed, made,** and **sold** all the components down to the light bulb,
and began to **export their system** throughout the U.S. and to countries **around the world.**

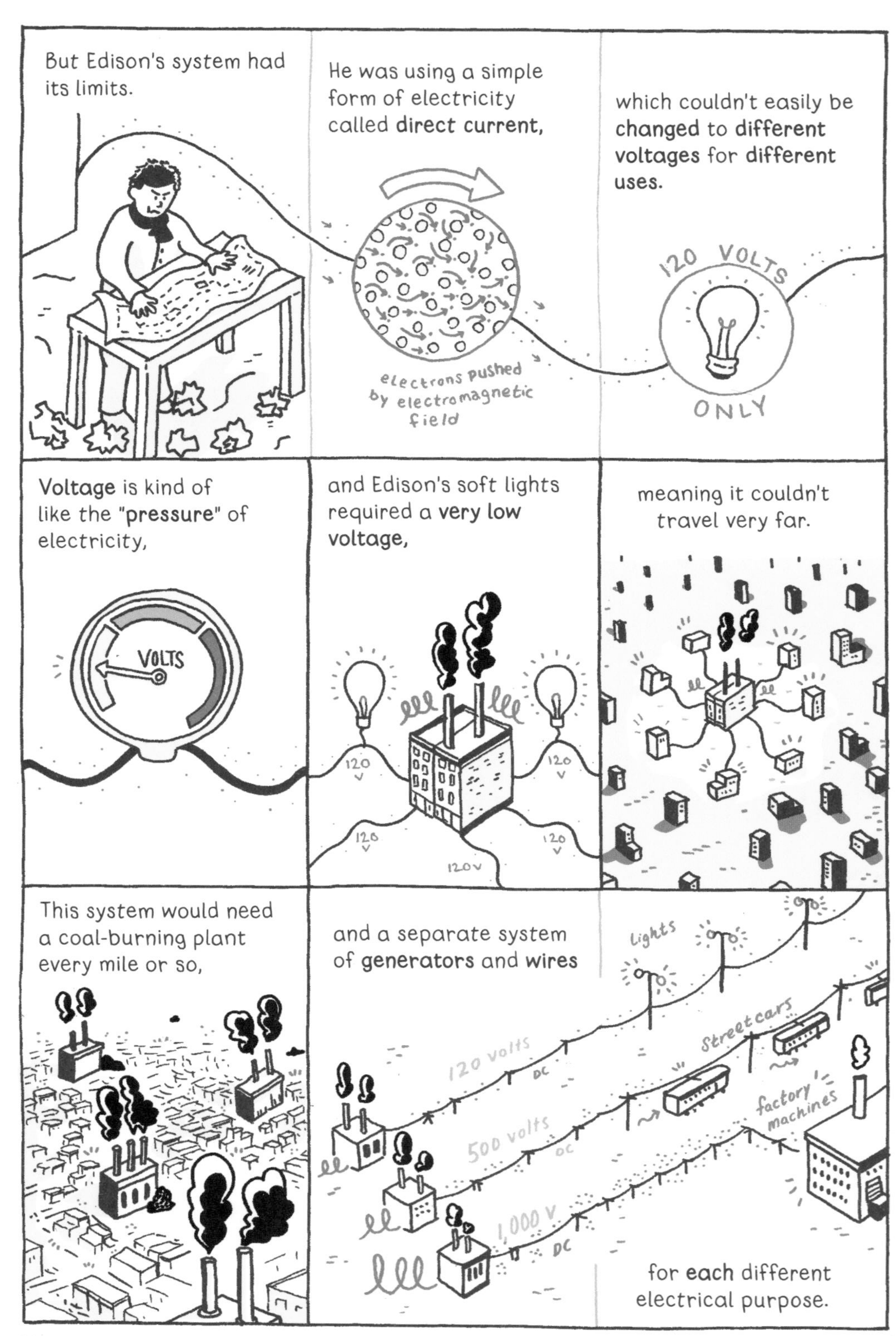
But Edison's system had its limits.
He was using a simple form of electricity called **direct current,**
electrons pushed by electromagnetic field
which couldn't easily be **changed** to **different voltages** for **different uses.**
120 VOLTS ONLY
Voltage is kind of like the "**pressure**" of electricity,
VOLTS
and Edison's soft lights required a **very low voltage,**
120 V
120 V
120 V
120 V
120V
meaning it couldn't travel very far.
This system would need a coal-burning plant every mile or so,
and a separate system of **generators** and **wires**
Lights
120 volts DC
Streetcars
500 volts DC
factory machines
1,000 V DC
for **each** different electrical purpose.

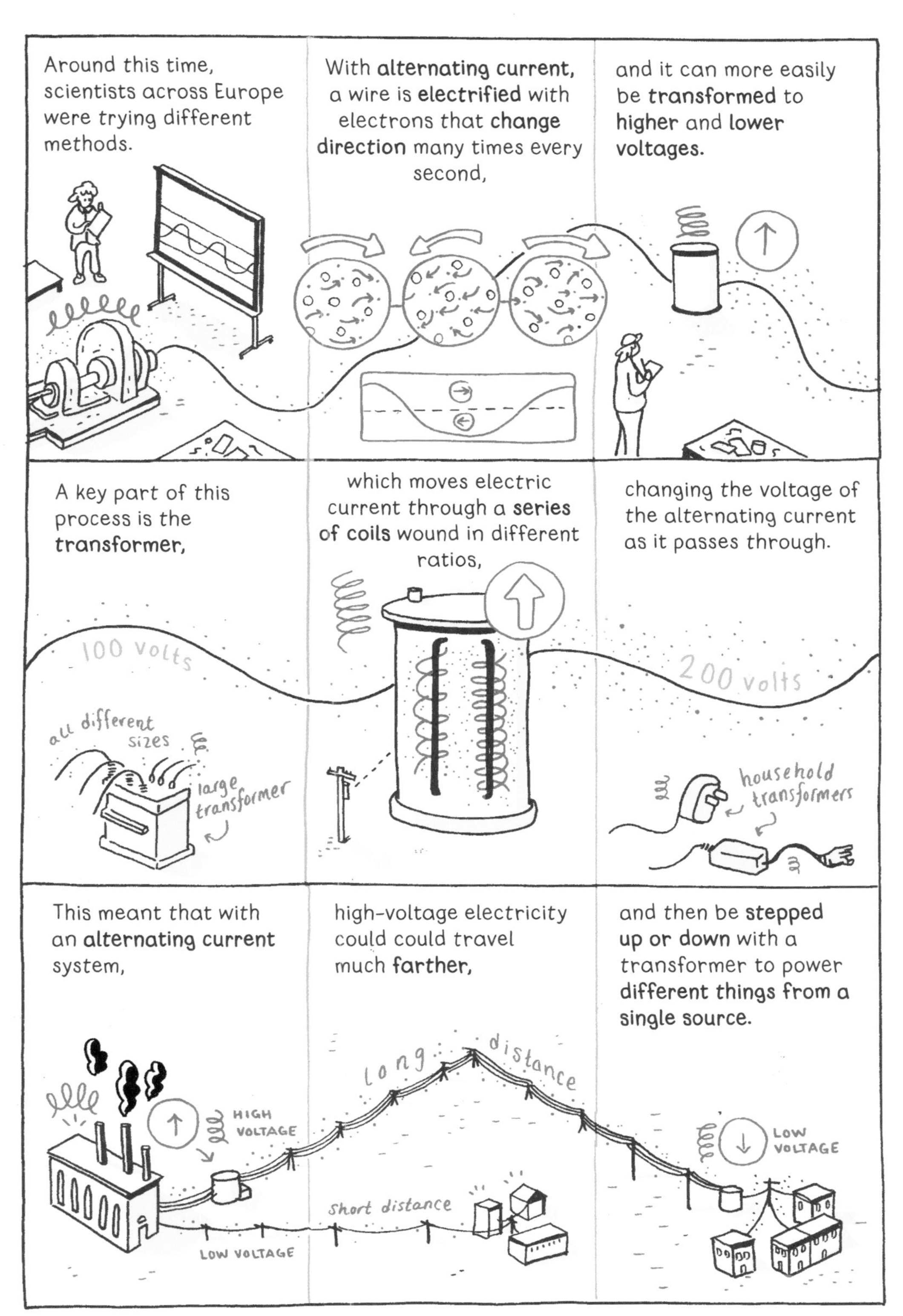
Around this time, scientists across Europe were trying different methods.
With **alternating current,** a wire is **electrified** with electrons that **change direction** many times every second,
and it can more easily be **transformed** to **higher** and **lower voltages.**
A key part of this process is the **transformer,**
which moves electric current through a **series of coils** wound in different ratios,
changing the voltage of the alternating current as it passes through.
100 volts
all different sizes
large transformer
200 volts
household transformers
This meant that with an **alternating current** system,
high-voltage electricity could could travel much **farther,**
and then be **stepped up or down** with a transformer to power **different things from a single source.**
HIGH VOLTAGE
LOW VOLTAGE
long distance
short distance
LOW VOLTAGE

Nikola Tesla,
a Serbian immigrant with a brilliant scientific imagination,
very well dressed
very eccentric
saw enormous, world-changing potential in **alternating current (AC)**
AC
and **created a motor** that could turn using this tricky oscillating electricity.
electric energy
magnets
to mechanical energy
George Westinghouse,
a businessman and an inventor himself,
also saw potential in alternating current
and teamed with Tesla to bring it to market.
shout-out to the laborers who did most of this work.
Edison would not admit that others had developed a better system.
He ran a **smear campaign** against alternating current, and used it to kill animals to show its danger.
!
!
uh wut
Edison even helped invent the **first electric chair** and attempted to associate AC with death.
It should use the alternating current of Westinghouse.

Backers of both systems wanted to show off, and competed on big **industry-defining projects,**

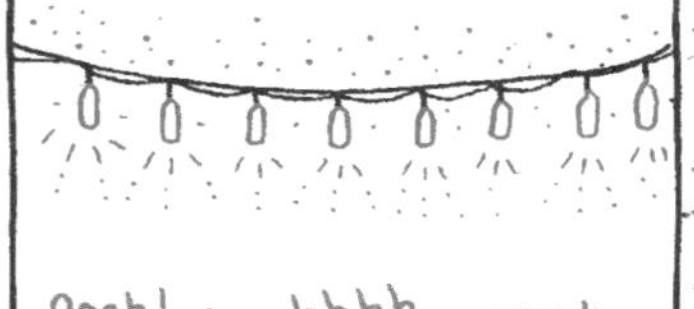

like providing light for the **1893 World's Fair in Chicago**

and building generators to **harness the energy of Niagara Falls.**

Alternating current proved more versatile and is **mostly what we use today.**

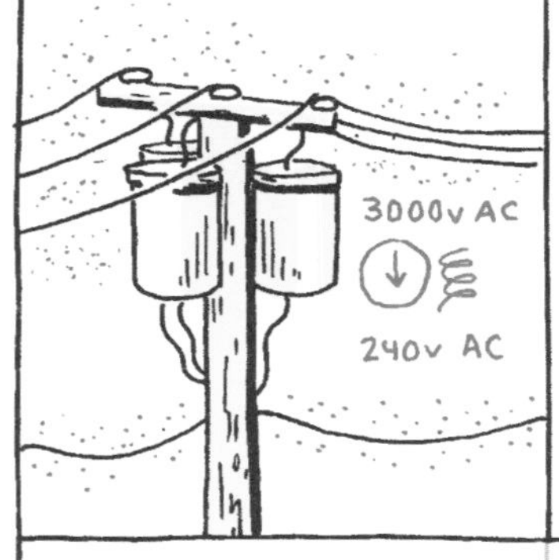

Edison's company was sold from under him by investors and renamed,

and he moved on to other endeavors, including trying to monopolize a new and growing industry.

Tesla went back to the thing he loved most—**inventing.**

He worked toward his dream of building a **global network of wireless electricity and information.**

But in the **early 1900s,** it was a concept he got little funding for.

clink
clink
clank

Building the Grid: Transmission

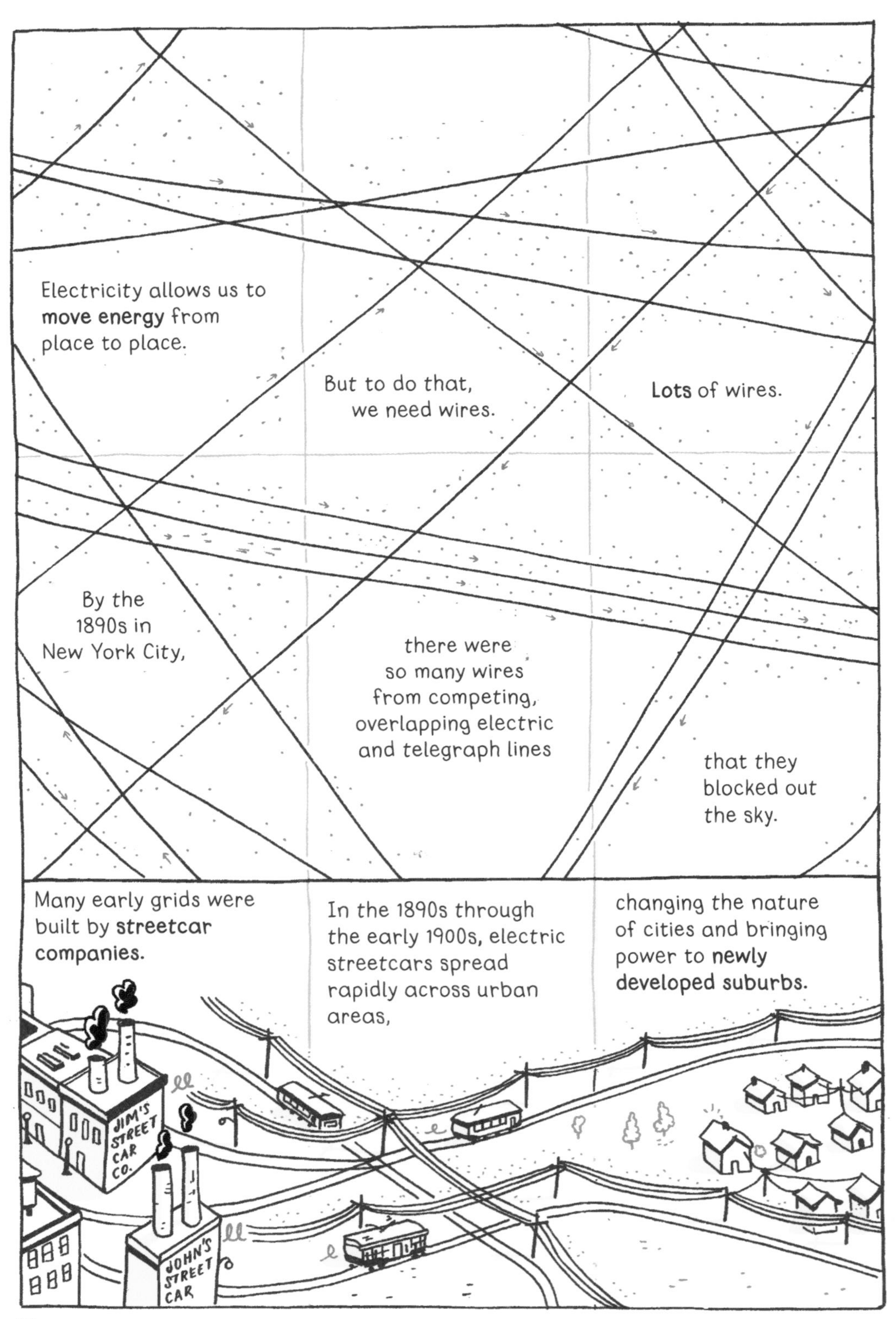
Electricity allows us to **move energy** from place to place.
But to do that, we need wires.
Lots of wires.
By the 1890s in New York City,
there were so many wires from competing, overlapping electric and telegraph lines
that they blocked out the sky.
Many early grids were built by **streetcar companies.**
In the 1890s through the early 1900s, electric streetcars spread rapidly across urban areas,
changing the nature of cities and bringing power to **newly developed suburbs.**
JIM'S STREET CAR CO.
JOHN'S STREET CAR

Interest in electricity and its uses was **growing**, but it was still a **niche technology**.

electric cars 1890s

steam car

animal car

gas car

Aside from a few "**central stations**" like Pearl Street,

most electricity was produced **where it was used.**

In many countries, electricity was emerging as a **public good provided by the government.**

Isn't it nice when the government does something useful?

In the **U.S.**, it was a **commodity** that was developed **privately,** both by **businesses**

and by the **wealthy,** who often installed mini power plants to light their homes.

At the turn of the 20th century, electricity was still a **mysterious, futuristic force,**

written about by poets like Emily Dickinson

which, compared to steam power and gas, represented **progress and modernity,**

has never seen electricity

ooooh

ahhhh

but had not yet been fully developed into a product for use by the **general public.**

hm there must be SOME way to sell this.

Someone who did see the **commercial potential** for electricity was **Samuel Insull.**

Starting at Edison's small Chicago branch, he figured that the **more customers he could sell to,**

the cheaper he could provide electricity for,

applying a concept called "**economies of scale**" to energy.

Insull built **huge generators,**

sold electricity **cheaply,**

and convinced others to **buy his power,** rather than **produce it themselves.**

With various types of customers needing power at different times of day, he had the demand to run the plants **around the clock.**

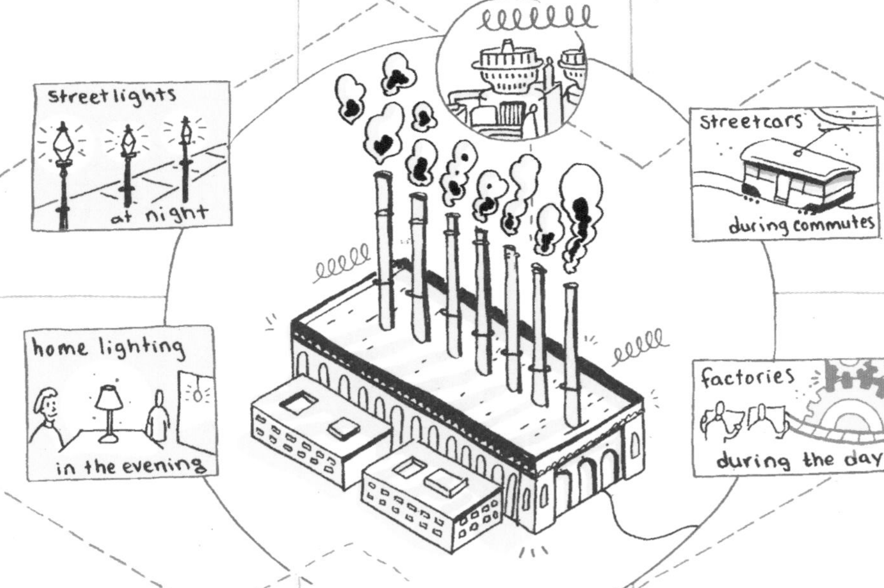

Insull bought out competing plants

and lobbied for electricity production to be a kind of **regulated monopoly** called a **public utility.**

In exchange for government-set rates, Insull was guaranteed **no competition** in his region.

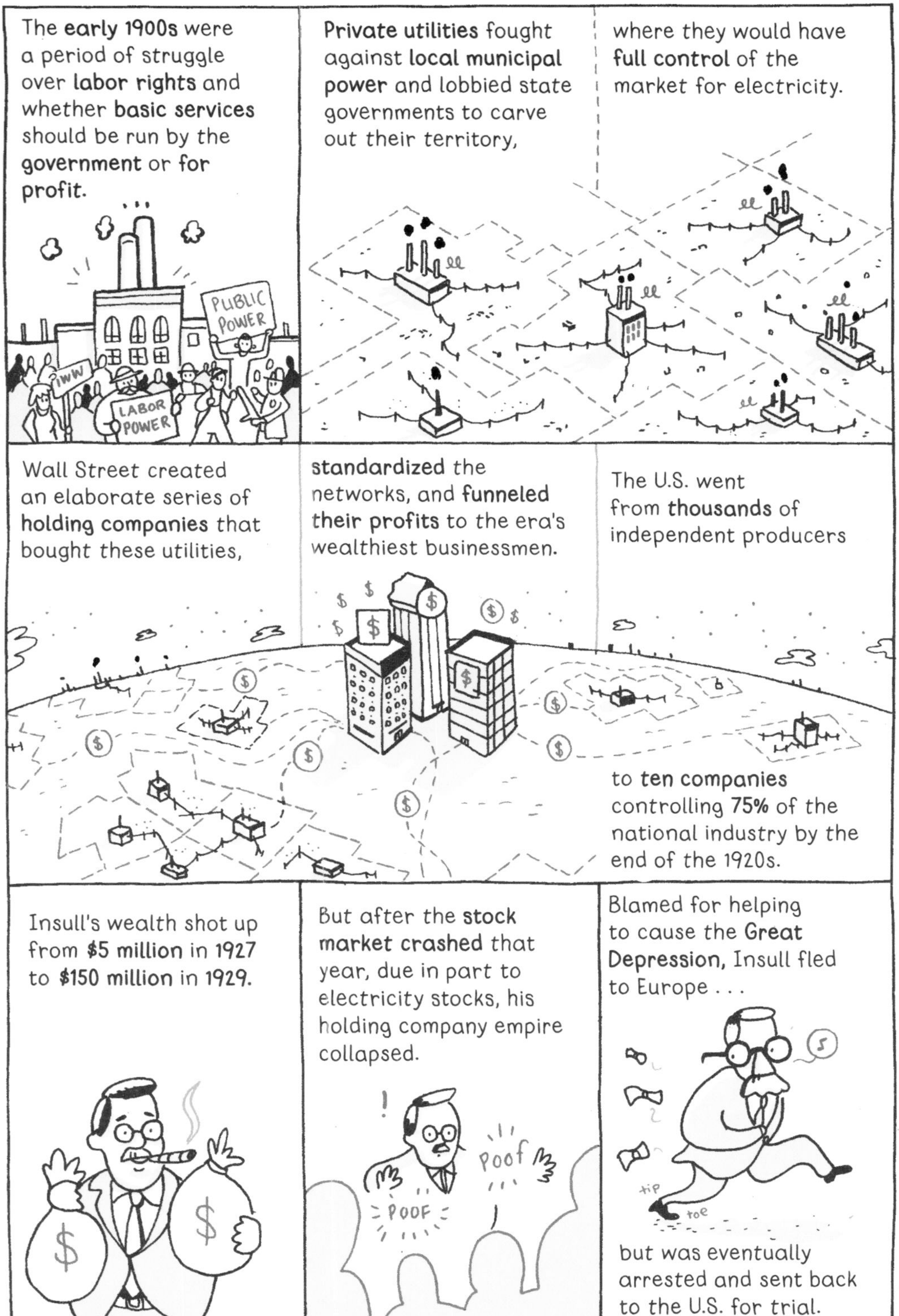

The **early 1900s** were a period of struggle over **labor rights** and whether **basic services** should be run by the **government** or **for profit.**

Private utilities fought against **local municipal power** and lobbied state governments to carve out their territory,

where they would have **full control** of the market for electricity.

Wall Street created an elaborate series of **holding companies** that bought these utilities,

standardized the networks, and **funneled their profits** to the era's wealthiest businessmen.

The U.S. went from **thousands** of independent producers

to **ten companies** controlling **75%** of the national industry by the end of the 1920s.

Insull's wealth shot up from **$5 million** in 1927 to **$150 million** in 1929.

But after the **stock market crashed** that year, due in part to electricity stocks, his holding company empire collapsed.

Blamed for helping to cause the **Great Depression,** Insull fled to Europe . . .

but was eventually arrested and sent back to the U.S. for trial.

At the same time,
as late as the **1930s**, much of the rural U.S. was still **without electricity**
because private utilities claimed that it would be too expensive to serve them power.
burning a kerosene lamp
In the early 1900s, the divide between **rural** and **urban** existence was stark.
Without electricity, farm life was grueling,
and it was made worse by the poverty of the Great Depression.
wash
wash
wash
clean
clean

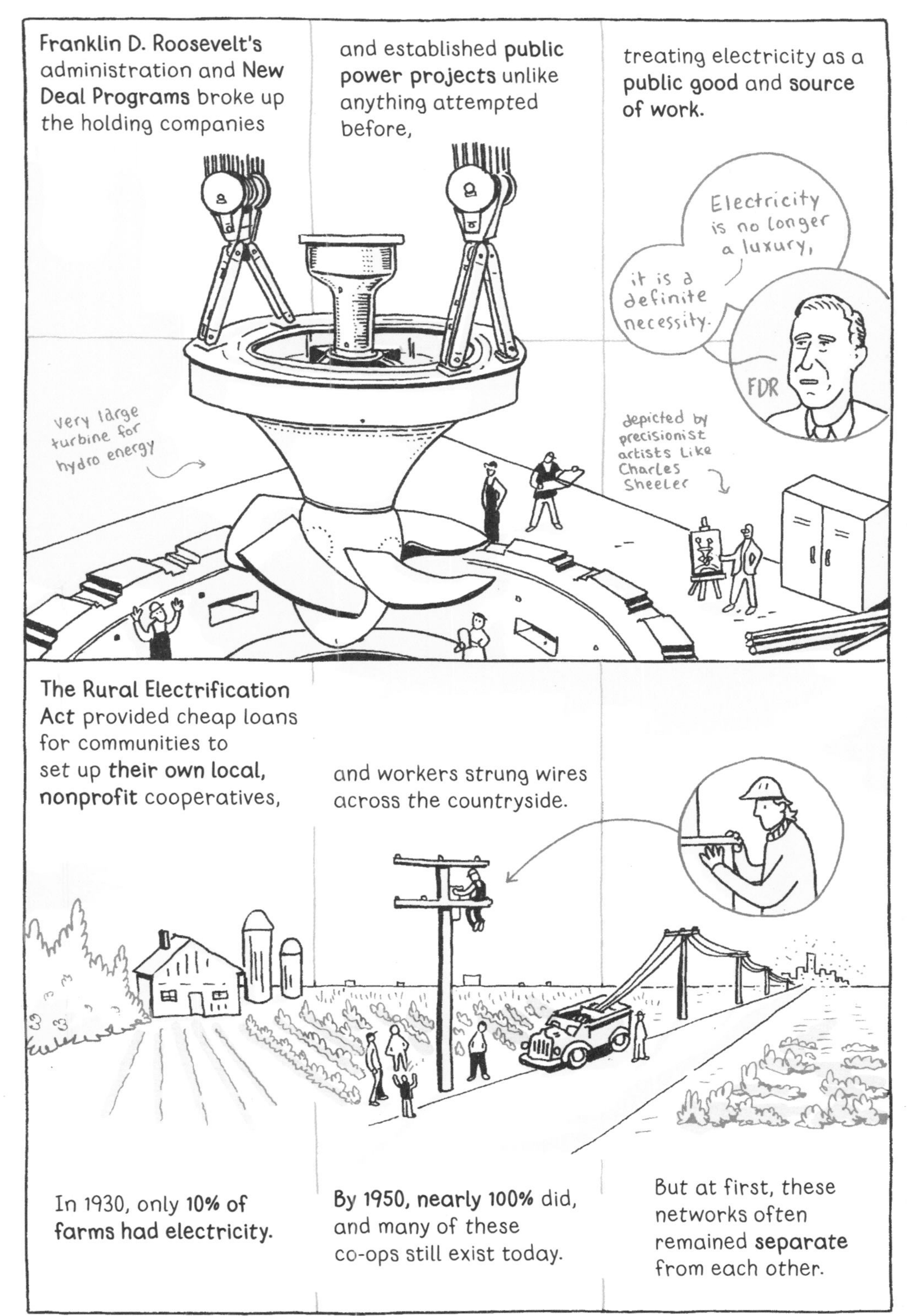
Franklin D. Roosevelt's administration and New Deal Programs broke up the holding companies
and established public power projects unlike anything attempted before,
treating electricity as a public good and source of work.
Electricity is no longer a luxury,
it is a definite necessity.
FDR
very large turbine for hydro energy
depicted by precisionist artists like Charles Sheeler
The Rural Electrification Act provided cheap loans for communities to set up their own local, nonprofit cooperatives,
and workers strung wires across the countryside.
In 1930, only 10% of farms had electricity.
By 1950, nearly 100% did, and many of these co-ops still exist today.
But at first, these networks often remained separate from each other.

In 1939, when World War II broke out,
the United States needed to quickly **produce millions of tons of weapons and materials,**
but faced a serious potential **shortage of electricity.**
The war industry required more power than had ever been needed before.
~8,800 large ships
food + supplies
~297,000 planes
~86,000 tanks

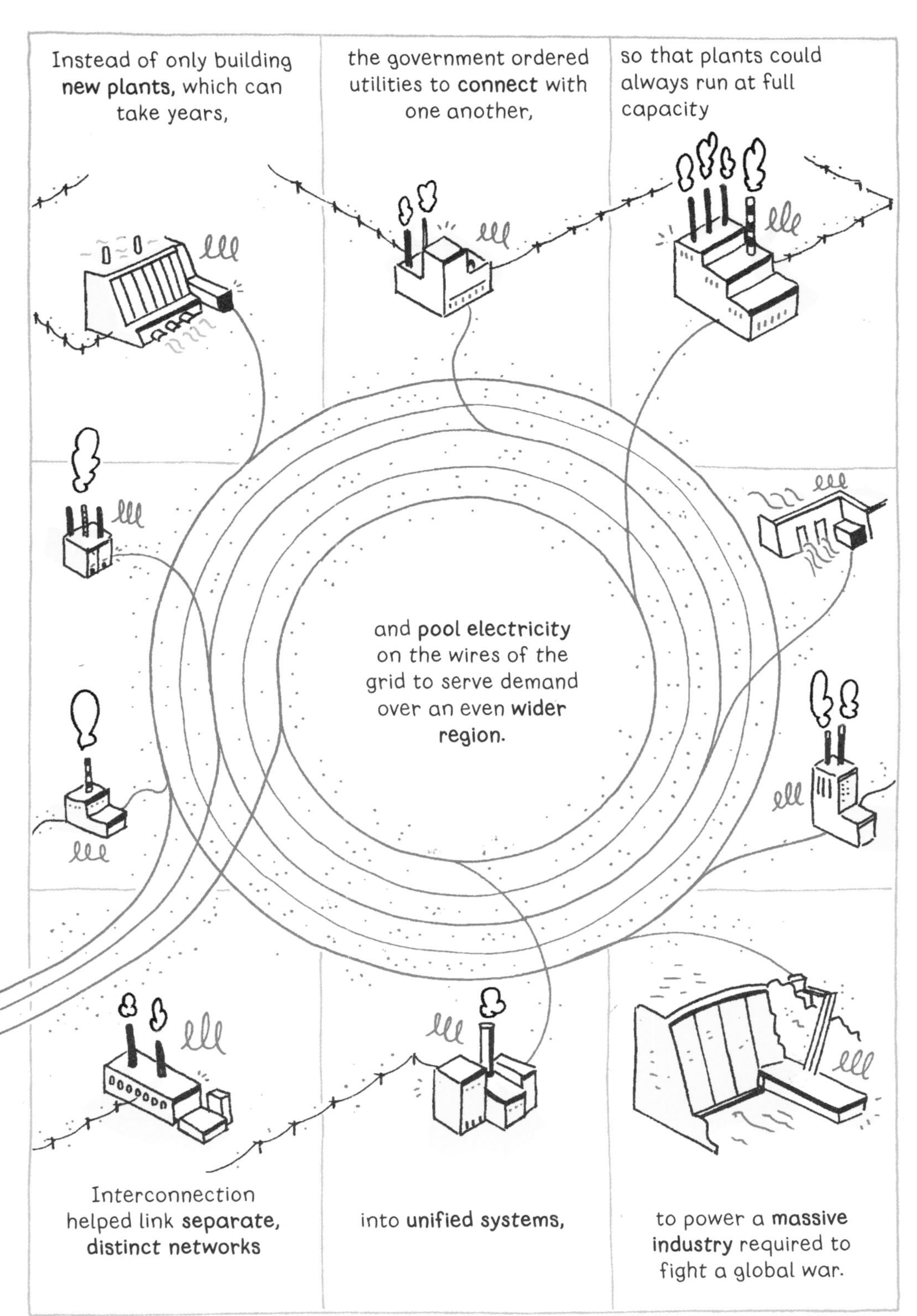
Instead of only building **new plants,** which can take years,
the government ordered utilities to **connect** with one another,
so that plants could always run at full capacity
and **pool electricity** on the wires of the grid to serve demand over an even **wider region.**
Interconnection helped link **separate, distinct networks**
into **unified systems,**
to power a **massive industry** required to fight a global war.

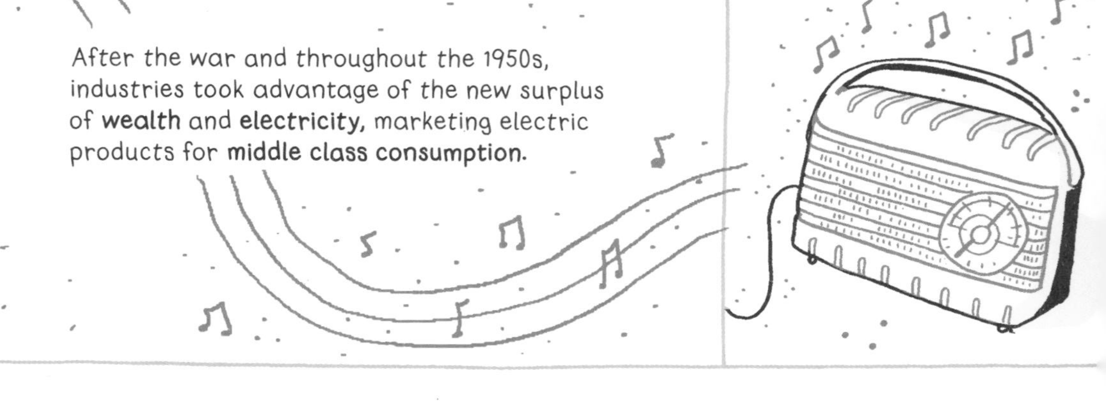

Electric utilities and companies like **General Electric** helped market a range of products that would use up **lots of power**,

often sending out reps to teach families how they could electrify their **homes** and **labor**.

By the 1960s, a majority of homes had a range of electrical devices, which became a standard of American **economy** and **culture**.

The grid that powers our homes and factories wasn't built with an **overarching plan;** it was **stitched together** by **companies** and **governments, piece by piece.**

Long-distance lines were built to move energy from places of **abundance** to places of **demand,** first powering **cities,**

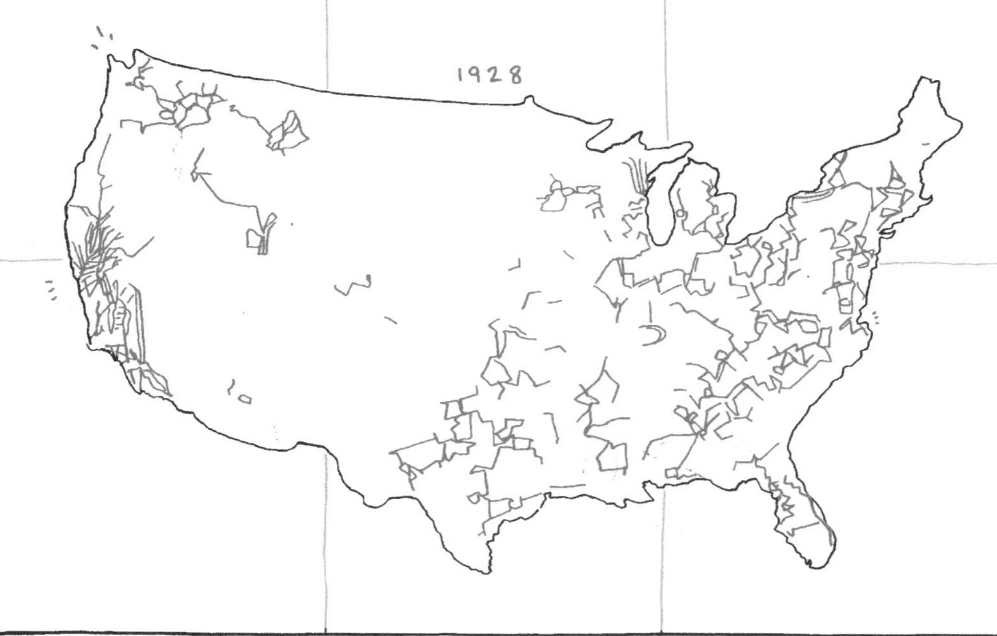

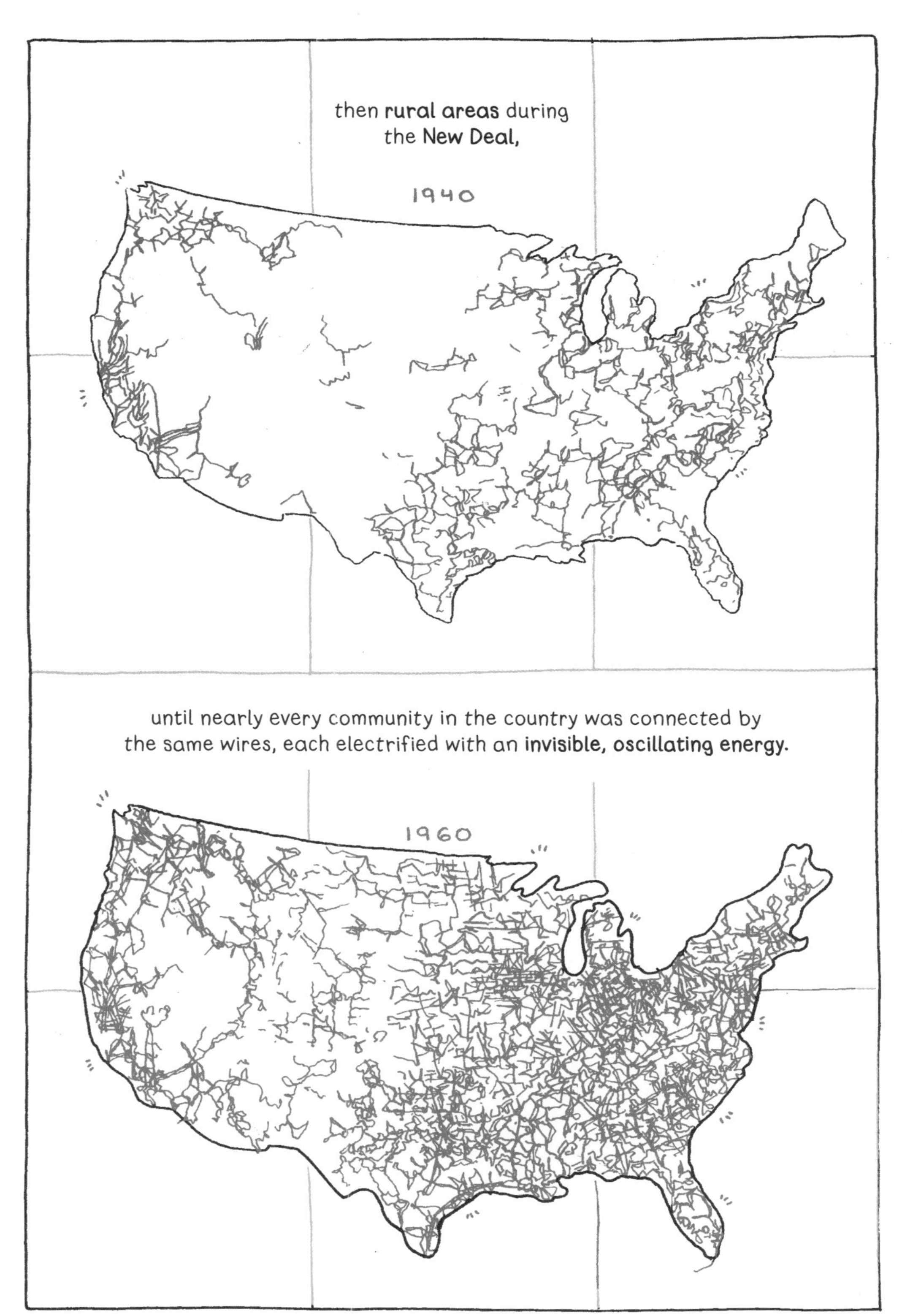
then **rural areas** during the **New Deal,**
1940
until nearly every community in the country was connected by the same wires, each electrified with an **invisible, oscillating energy.**
1960

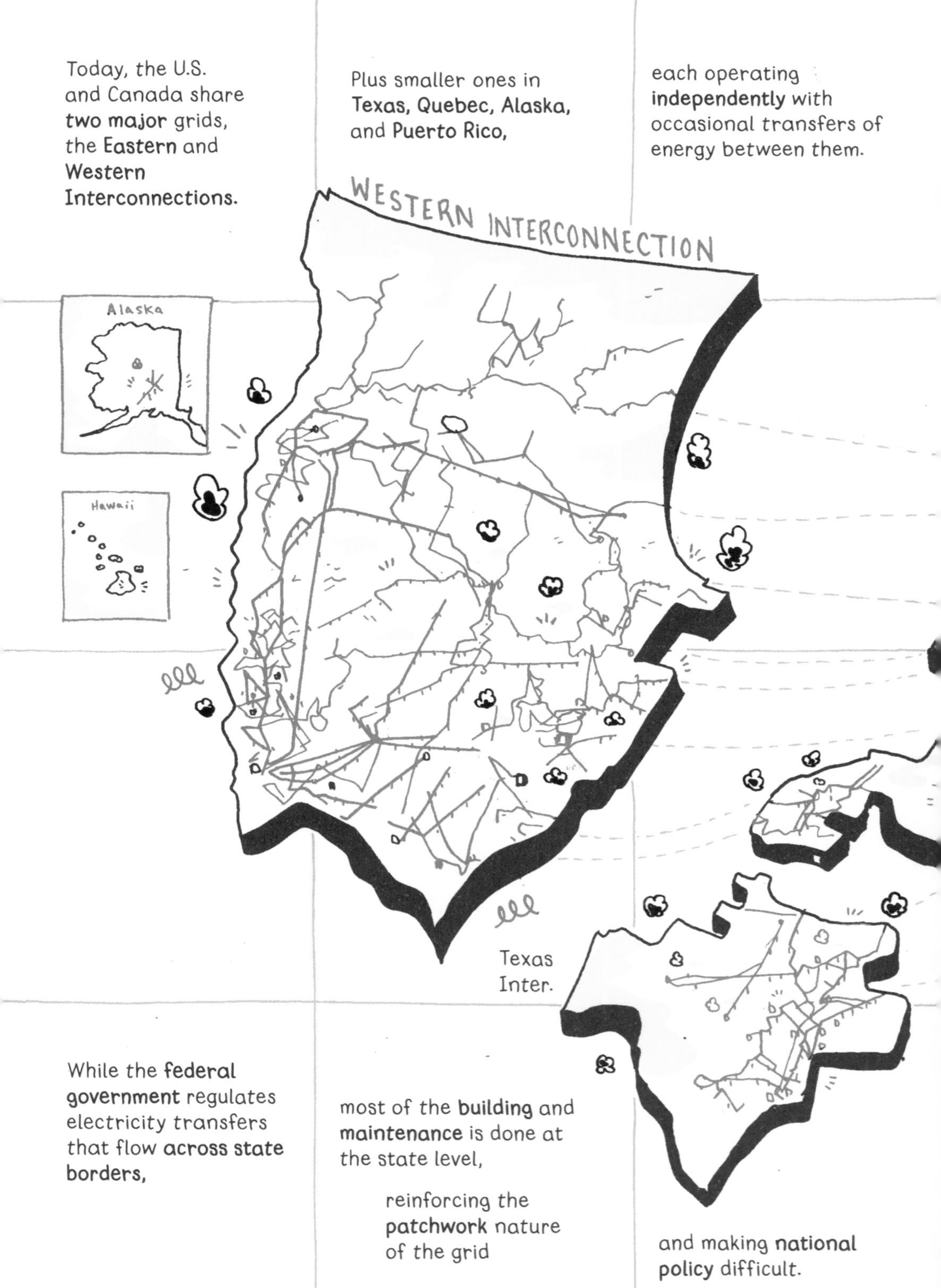
Today, the U.S. and Canada share **two major** grids, the **Eastern** and **Western Interconnections.**
Plus smaller ones in **Texas, Quebec, Alaska,** and **Puerto Rico,**
each operating **independently** with occasional transfers of energy between them.
WESTERN INTERCONNECTION
Alaska
Hawaii
Texas Inter.
While the **federal government** regulates electricity transfers that flow **across state borders,**
most of the **building** and **maintenance** is done at the state level,
reinforcing the **patchwork** nature of the grid
and making **national policy** difficult.

In each of these grids, from top to bottom, electricity **flows freely.**

The electricity in every plant and transmission line must be **perfectly synchronized by grid technicians,**

with the alternating current changing directions in unison, **sixty times a second** from **Toronto** to **Miami.**

Sixty-six different regional organizations monitor and operate the local sections of their grid,

making sure that electricity is transmitted reliably across their area,

and that **enough electricity** is being **generated** from a variety of sources.

whew

Powering the Grid: Generation and Fuel

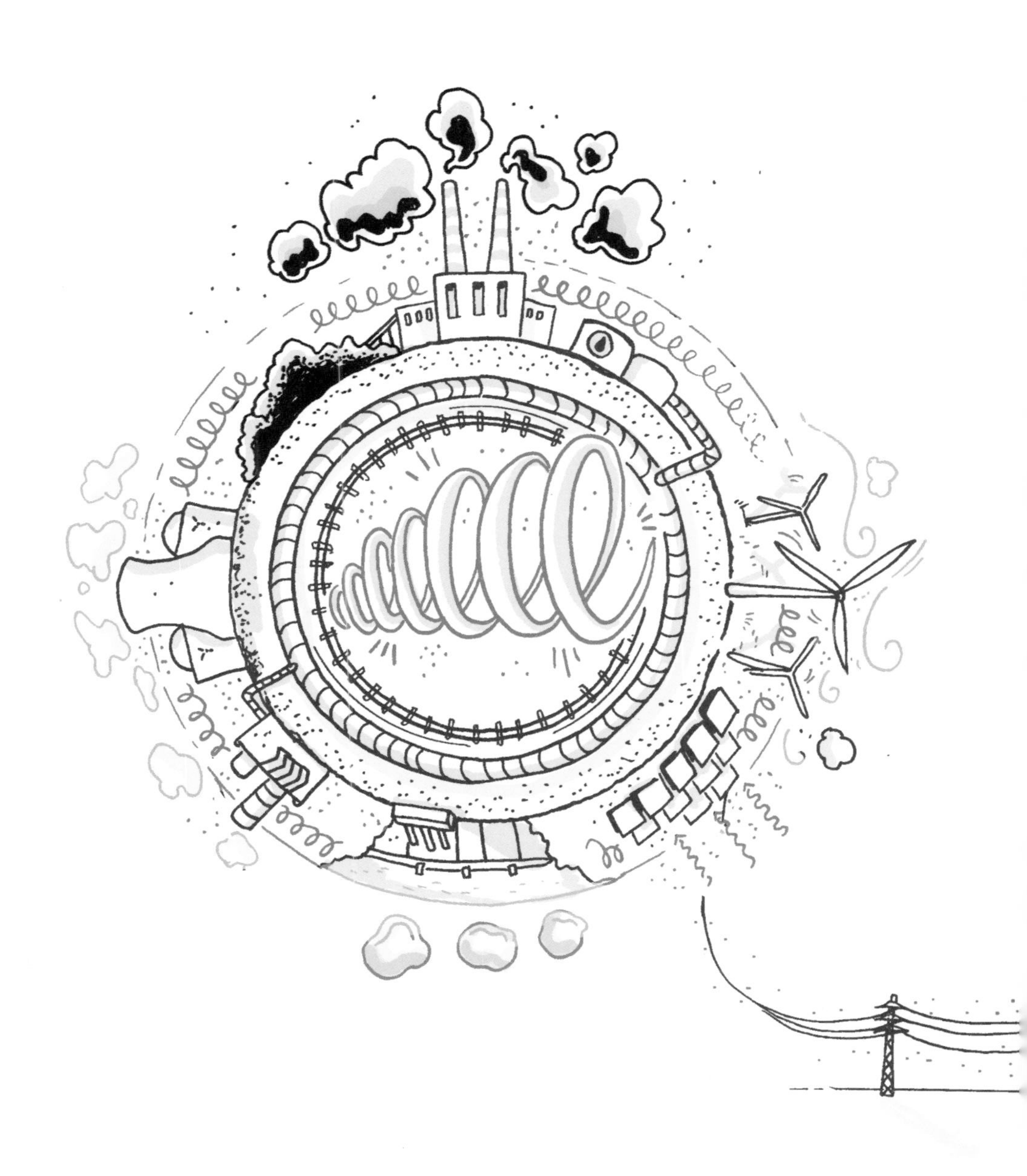

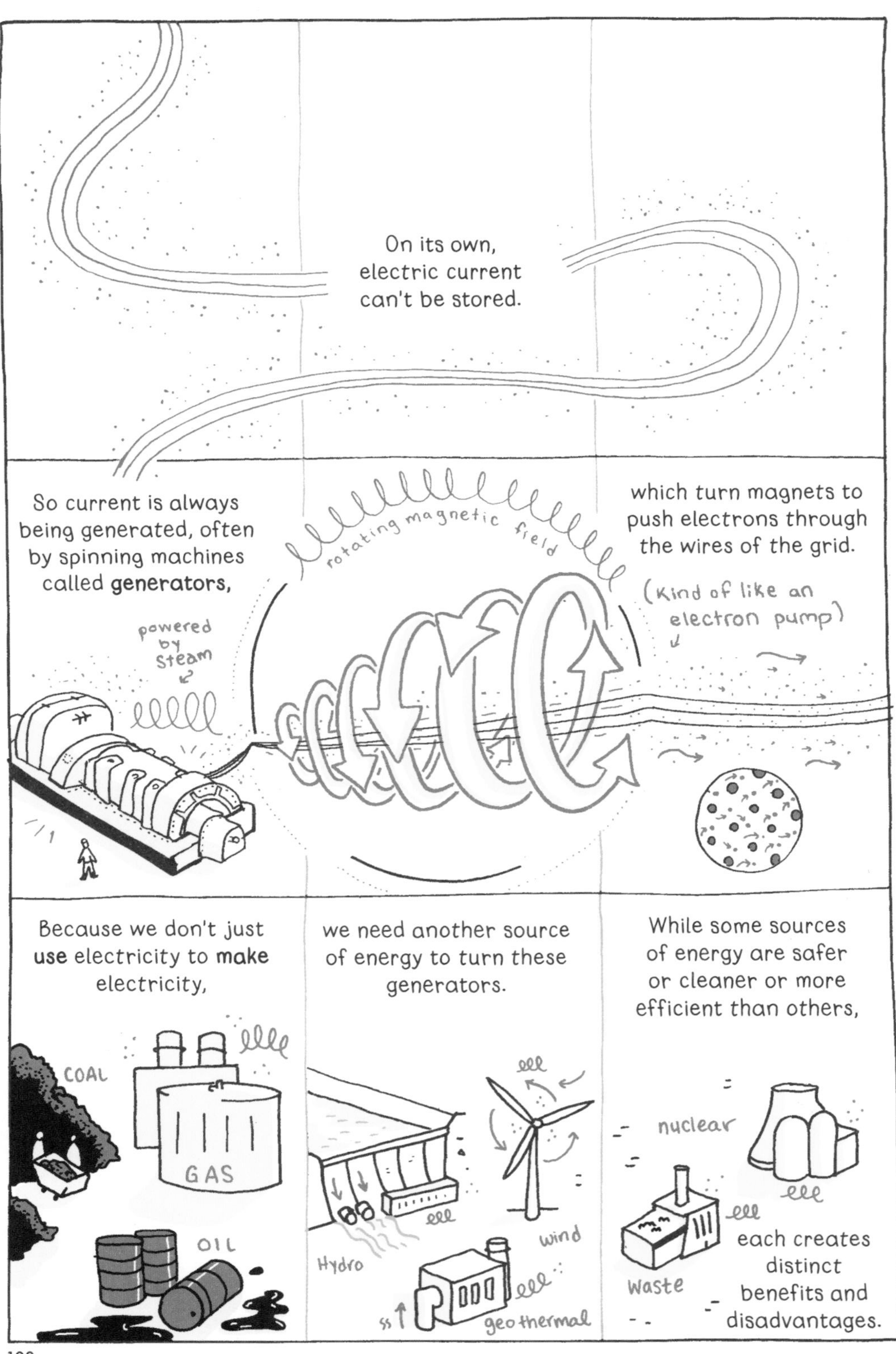
On its own, electric current can't be stored.
So current is always being generated, often by spinning machines called **generators**,
rotating magnetic field
powered by steam
which turn magnets to push electrons through the wires of the grid.
(Kind of like an electron pump)
Because we don't just **use** electricity to **make** electricity,
COAL
GAS
OIL
we need another source of energy to turn these generators.
Hydro
wind
geothermal
While some sources of energy are safer or cleaner or more efficient than others,
nuclear
waste
each creates distinct benefits and disadvantages.

Early electrical devices often ran on **batteries,**
stores energy chemically
which use a chemical reaction between a mix of materials to generate current.
COPPER
electrolyte
ZINC
But for anything high-powered, you needed a **lot of** them.
hm. not ideal.
At the Pearl Street Station, Edison used **six giant generators** to provide the current to thousands of light bulbs,
current
spin
spin
spin
spin
spin
and to spin those generators,
he used **boilers and steam engines,**
STEAM
HEAT
powered by the round-the-clock burning of **thousands of pounds of coal.**

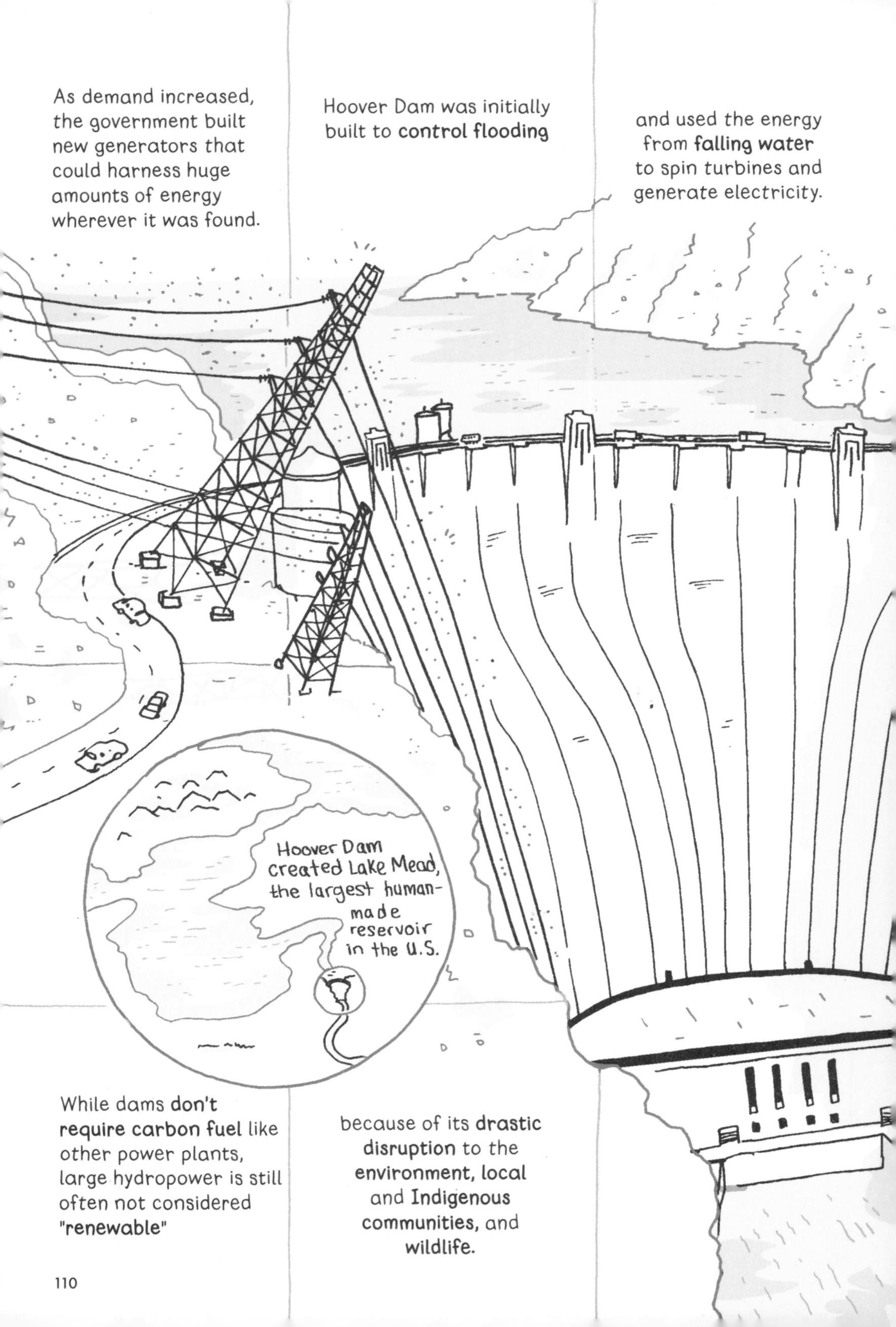
As demand increased, the government built new generators that could harness huge amounts of energy wherever it was found.
Hoover Dam was initially built to **control flooding**
and used the energy from **falling water** to spin turbines and generate electricity.
Hoover Dam created Lake Mead, the largest human-made reservoir in the U.S.
While dams **don't require carbon fuel** like other power plants, large hydropower is still often not considered "**renewable**"
because of its **drastic disruption** to the **environment, local** and **Indigenous communities,** and **wildlife.**

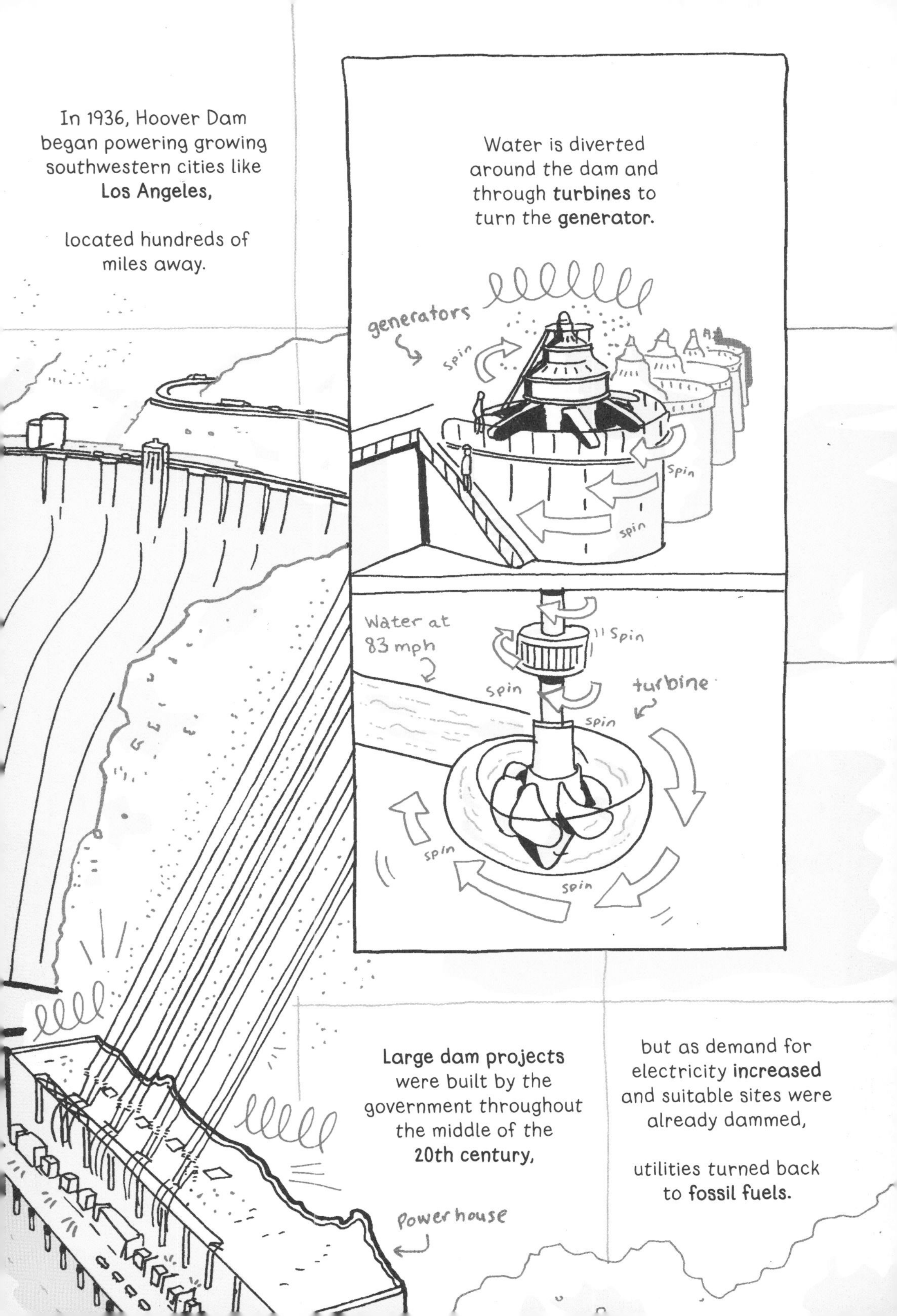
In 1936, Hoover Dam began powering growing southwestern cities like **Los Angeles,**
located hundreds of miles away.
Water is diverted around the dam and through **turbines** to turn the **generator.**
generators
spin
spin
spin
Water at 83 mph
spin
spin
turbine
spin
spin
spin
Large dam projects were built by the government throughout the middle of the **20th century,**
but as demand for electricity **increased** and suitable sites were already dammed,
utilities turned back to **fossil fuels.**
Powerhouse

Fossil fuels, like coal, oil, and natural gas,
Photosynthesis creates carbon.
are basically ancient, dead plants and life-forms that have been buried and condensed underground,
squish
but still retain the sun's energy.
squish
Because it burns well, people have mined coal for thousands of years.
CLANG
Because it's the planet's most abundant fossil fuel, it's easy and cheap to exploit for power.
Today it's extracted on a massive scale by companies and governments.
Coal is usually transported by rail,
BAM
coal trains
stored outside, and burned as needed for its heat,
water for steam
Powdered Coal
spinning a steam turbine and generator to produce electric current.
Burned for heat
creates steam
Turns
Turbine + generator

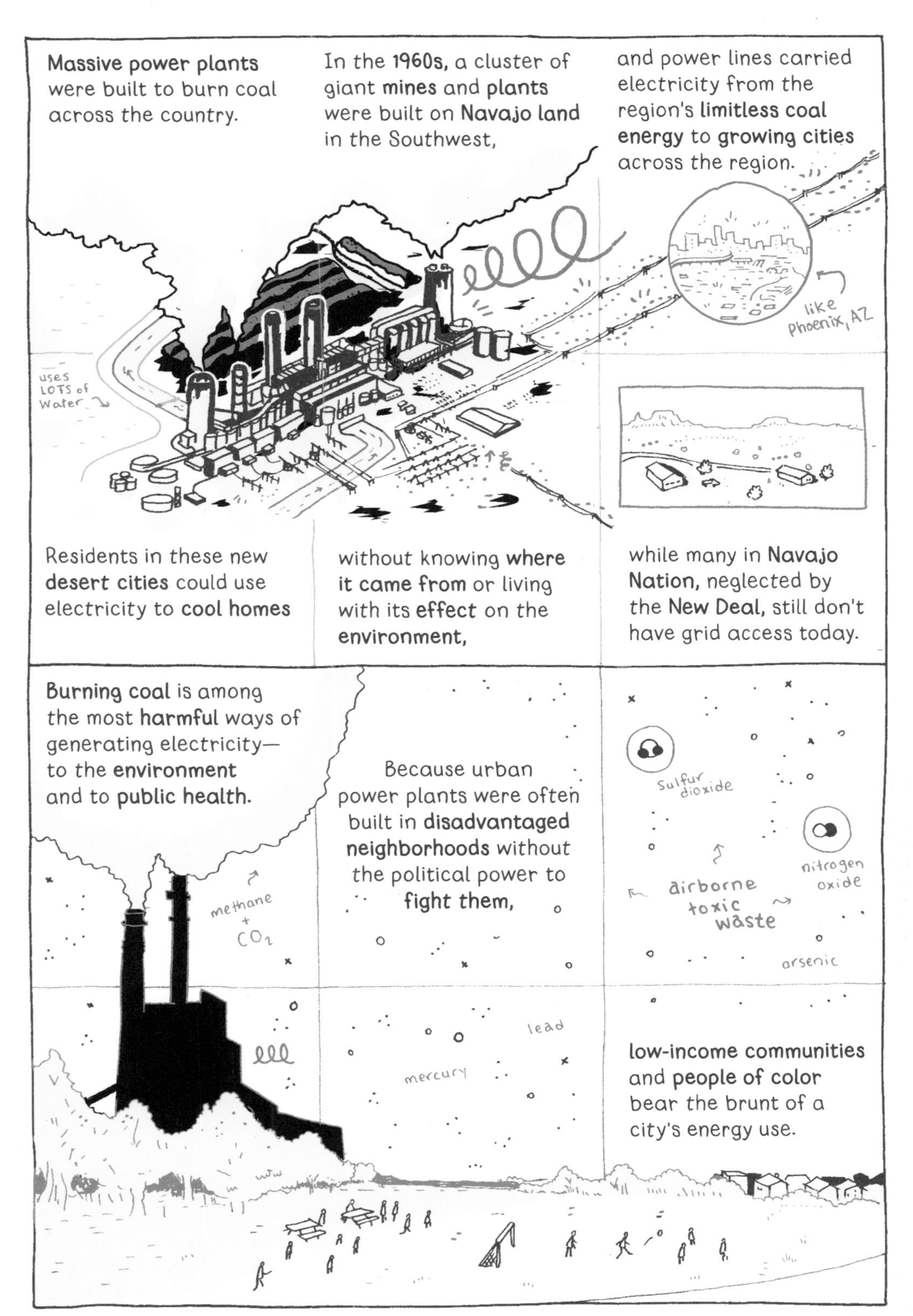

Massive power plants were built to burn coal across the country.
In the 1960s, a cluster of giant mines and plants were built on Navajo land in the Southwest,
and power lines carried electricity from the region's limitless coal energy to growing cities across the region.
uses LOTS of Water
like Phoenix, AZ
Residents in these new desert cities could use electricity to cool homes
without knowing where it came from or living with its effect on the environment,
while many in Navajo Nation, neglected by the New Deal, still don't have grid access today.
Burning coal is among the most harmful ways of generating electricity—to the environment and to public health.
methane + CO_2
Because urban power plants were often built in disadvantaged neighborhoods without the political power to fight them,
sulfur dioxide
nitrogen oxide
airborne toxic waste
arsenic
lead
mercury
low-income communities and people of color bear the brunt of a city's energy use.

Nuclear power plants were mostly built in the 1960s and '70s,
and despite being one of the **cleanest** ways to generate **lots** of electricity, **mining uranium** can be highly toxic,
and nuclear energy has never shaken its **associations** with the **atomic bomb** and **catastrophe.**
*plants can't actually explode like this, fyi.
water vapor
Cooling towers
nearly all reactors are made by G.E. and Westinghouse
3000 MW
reactors
Nuclear plants are also **super expensive** and **highly regulated,**
and a solution was never developed for the **long-term storage of the waste,**
so **new** plants are **rarely** built in the U.S.
These plants are basically a high-tech, low-emissions way of **heating water**—
uses pellets of enriched uranium as a fuel source
it's still all just to **create steam** to spin a turbine and generator.
reactor
steam
spin spin
turbine
generator
spin
Around **100 reactors** continue to provide about **20% of U.S. electricity** around the clock
and are used by over **thirty countries** worldwide.

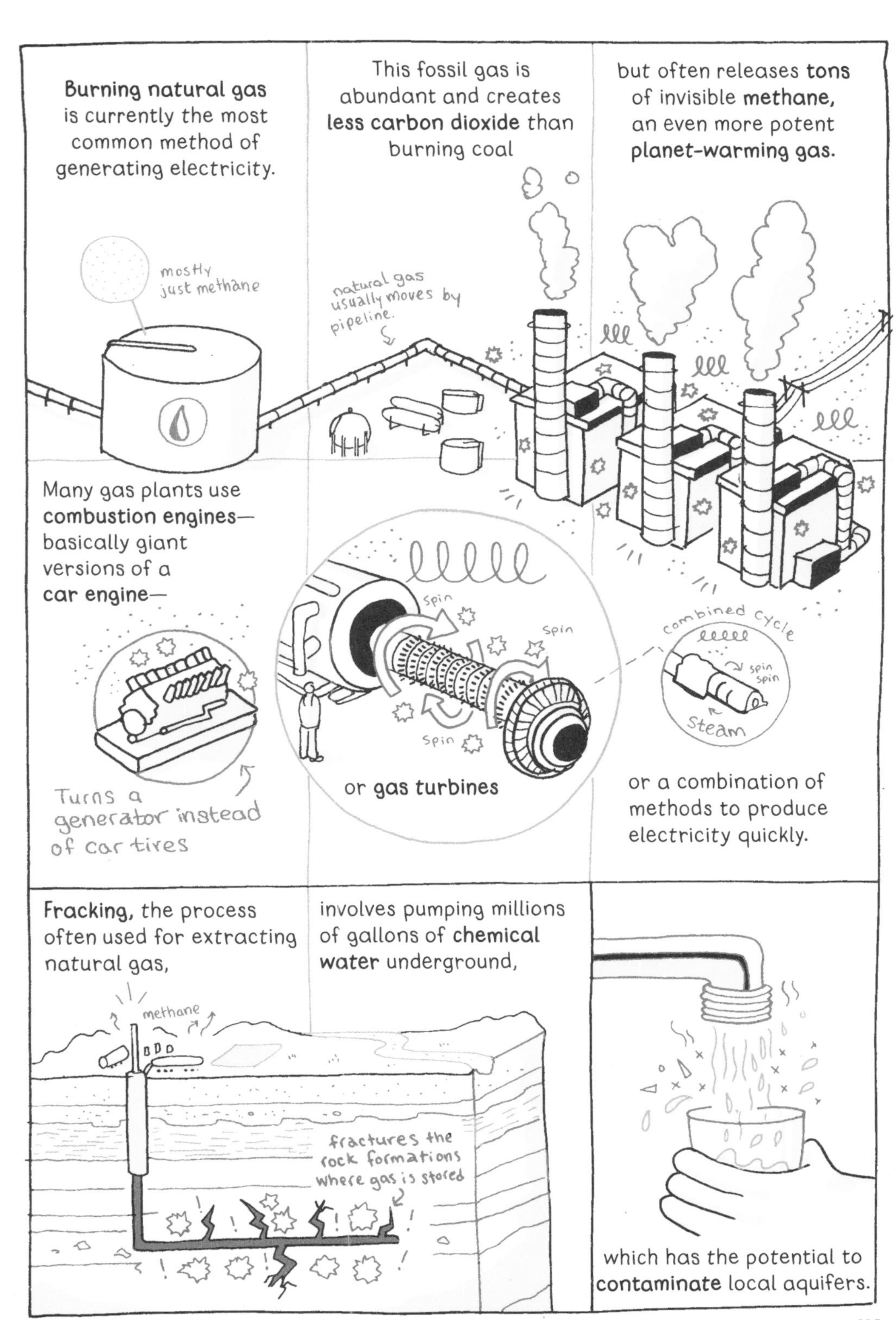

Burning natural gas is currently the most common method of generating electricity.
This fossil gas is abundant and creates **less carbon dioxide** than burning coal
but often releases **tons** of invisible **methane,** an even more potent **planet-warming gas.**
mostly just methane
natural gas usually moves by pipeline.
Many gas plants use **combustion engines**— basically giant versions of a **car engine**—
Turns a generator instead of car tires
spin
spin
spin
or **gas turbines**
combined cycle
spin spin
steam
or a combination of methods to produce electricity quickly.
Fracking, the process often used for extracting natural gas,
methane
involves pumping millions of gallons of **chemical water** underground,
fractures the rock formations where gas is stored
which has the potential to **contaminate** local aquifers.

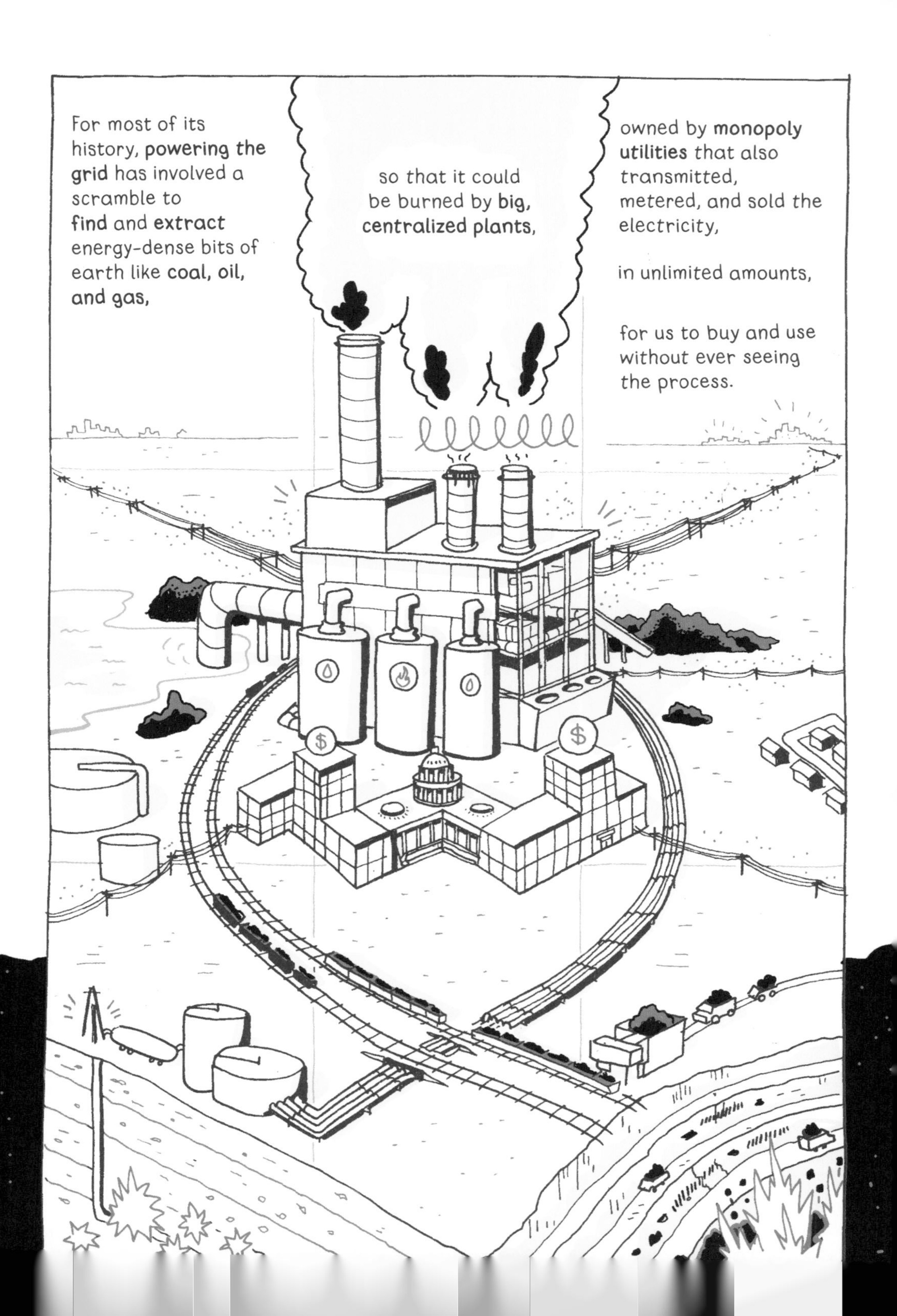
For most of its history, powering the grid has involved a scramble to find and extract energy-dense bits of earth like coal, oil, and gas,
so that it could be burned by big, centralized plants,
owned by monopoly utilities that also transmitted, metered, and sold the electricity,
in unlimited amounts,
for us to buy and use without ever seeing the process.

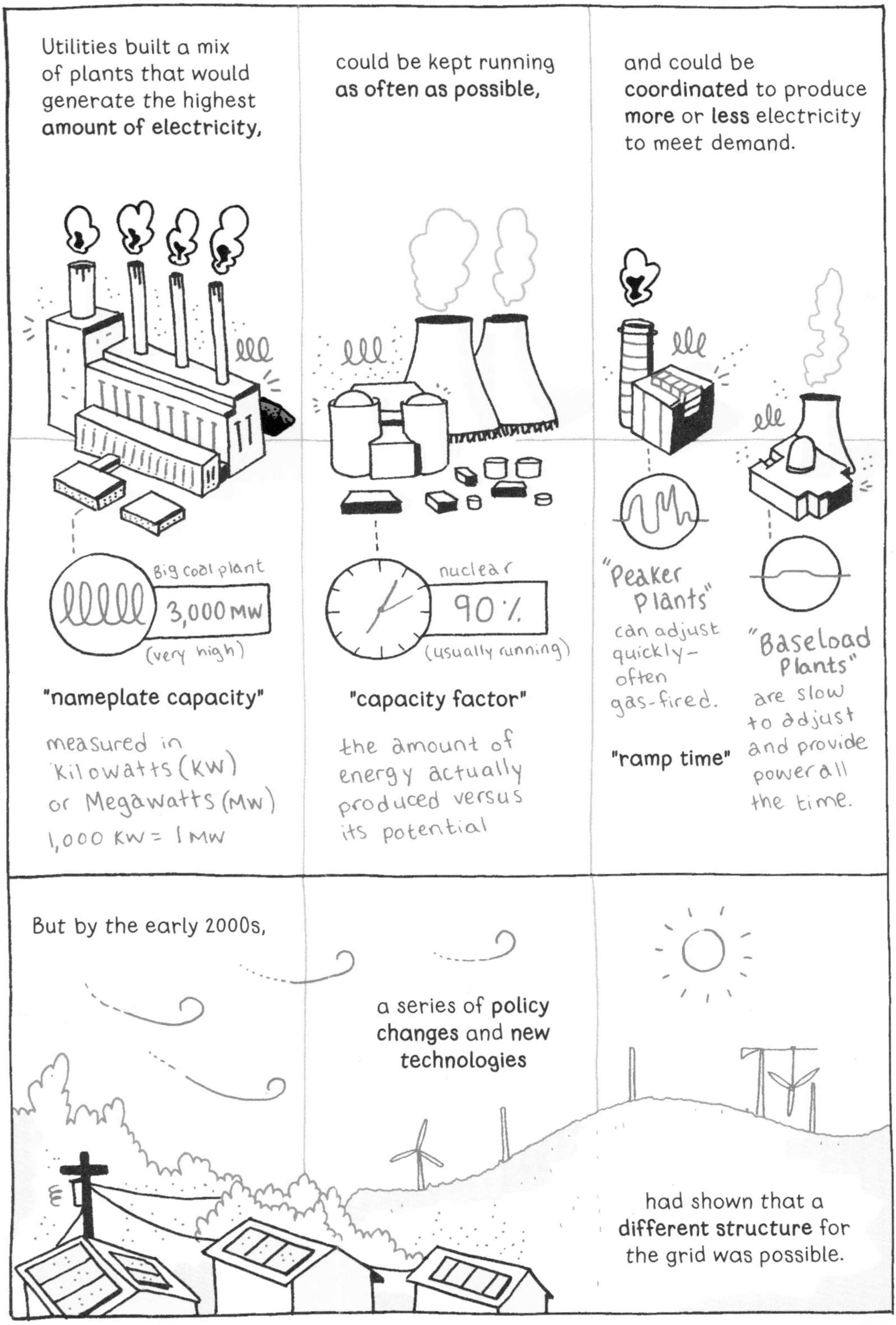

Utilities built a mix of plants that would generate the highest **amount of electricity,**

Big coal plant

3,000 MW

(very high)

"nameplate capacity"

measured in Kilowatts (KW) or Megawatts (MW)

1,000 KW = 1 MW

could be kept running **as often as possible,**

nuclear

90%

(usually running)

"capacity factor"

the amount of energy actually produced versus its potential

and could be **coordinated** to produce **more** or **less** electricity to meet demand.

"Peaker Plants" can adjust quickly – often gas-fired.

"ramp time"

"Baseload Plants" are slow to adjust and provide power all the time.

But by the early 2000s,

a series of **policy changes** and **new technologies**

had shown that a **different structure** for the grid was possible.

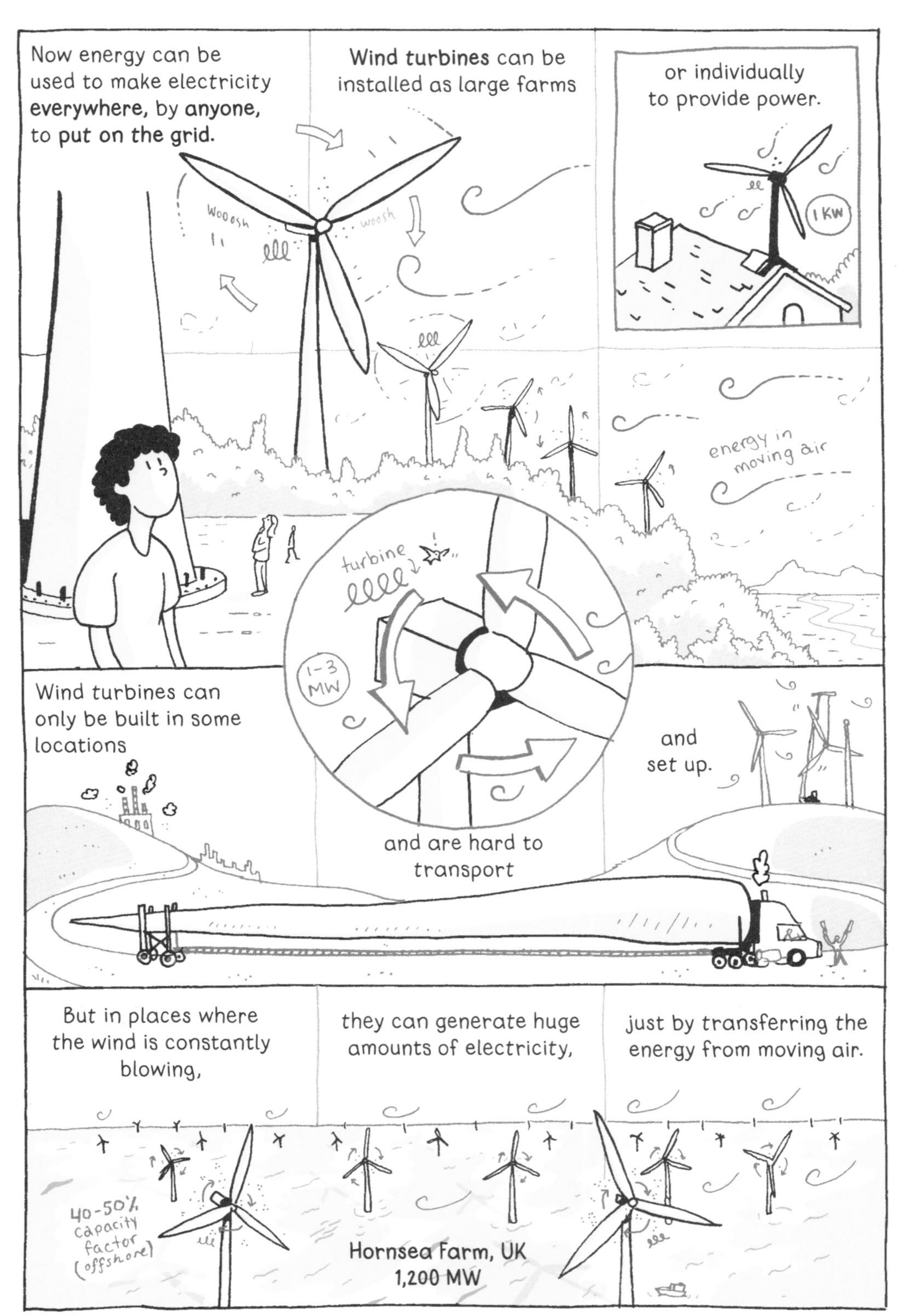

Now energy can be used to make electricity everywhere, by anyone, to put on the grid.
Wind turbines can be installed as large farms
or individually to provide power.
Wooosh
woosh
1 KW
energy in moving air
turbine
1-3 MW
Wind turbines can only be built in some locations
and are hard to transport
and set up.
But in places where the wind is constantly blowing,
they can generate huge amounts of electricity,
just by transferring the energy from moving air.
40-50% capacity factor (offshore)
Hornsea Farm, UK
1,200 MW

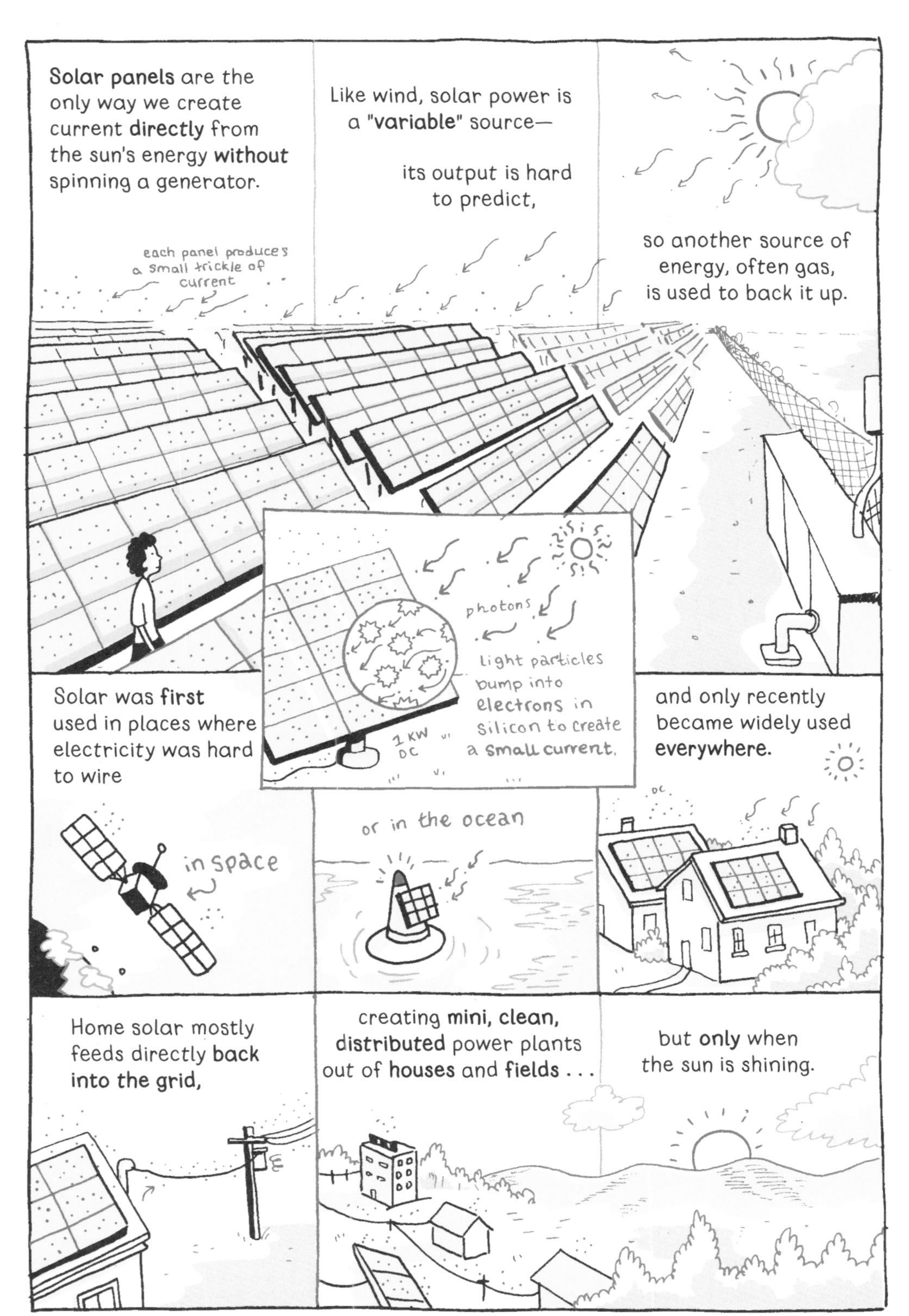
Solar panels are the only way we create current **directly** from the sun's energy **without** spinning a generator.
each panel produces a small trickle of current
Like wind, solar power is a "**variable**" source—
its output is hard to predict,
so another source of energy, often gas, is used to back it up.
photons
light particles bump into electrons in silicon to create a small current.
1 KW DC
Solar was **first** used in places where electricity was hard to wire
in space
or in the ocean
and only recently became widely used **everywhere.**
DC
Home solar mostly feeds directly **back into the grid,**
creating **mini, clean, distributed** power plants out of **houses** and **fields** . . .
but **only** when the sun is shining.

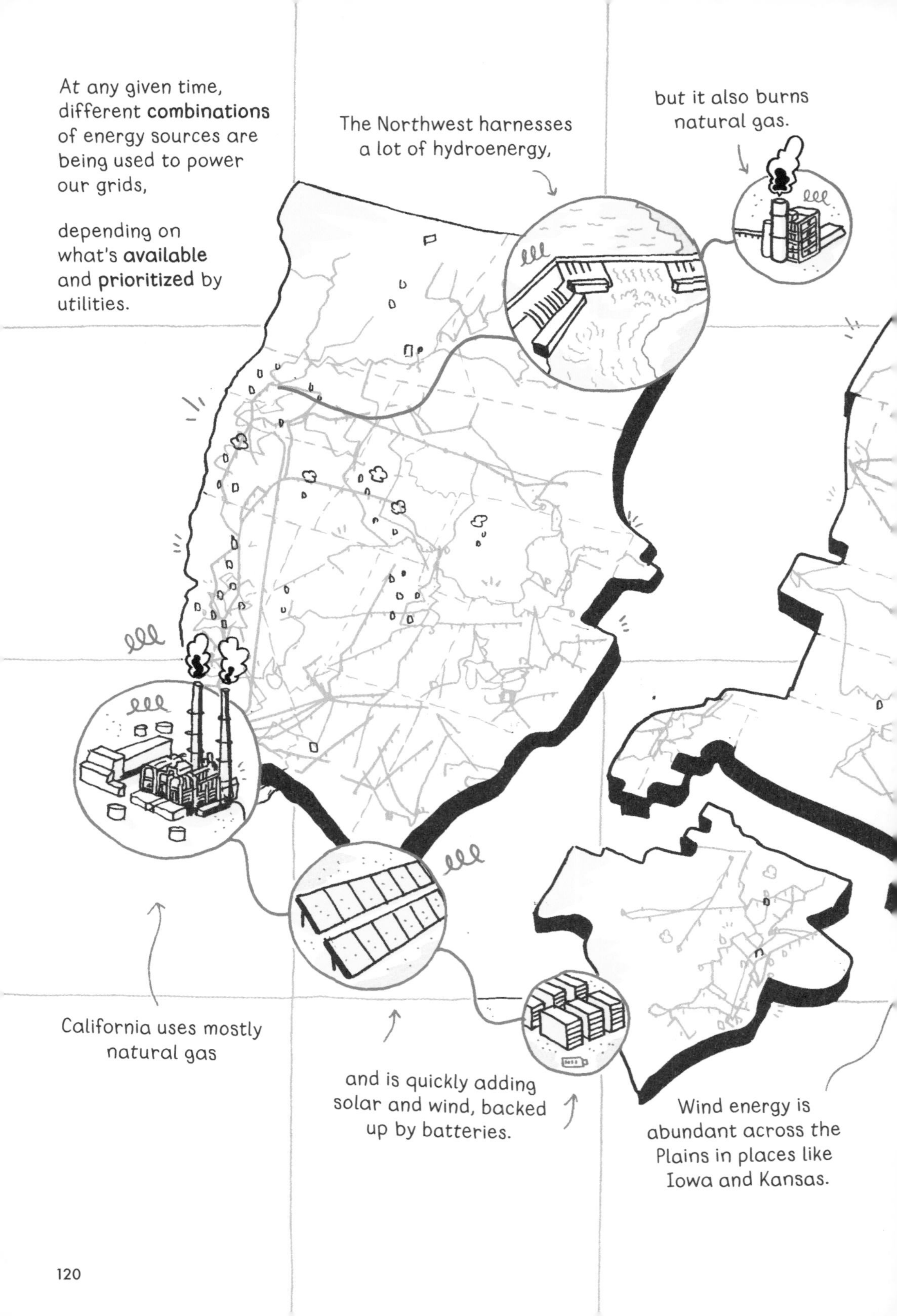
At any given time, different **combinations** of energy sources are being used to power our grids,
depending on what's **available** and **prioritized** by utilities.
The Northwest harnesses a lot of hydroenergy,
but it also burns natural gas.
California uses mostly natural gas
and is quickly adding solar and wind, backed up by batteries.
Wind energy is abundant across the Plains in places like Iowa and Kansas.

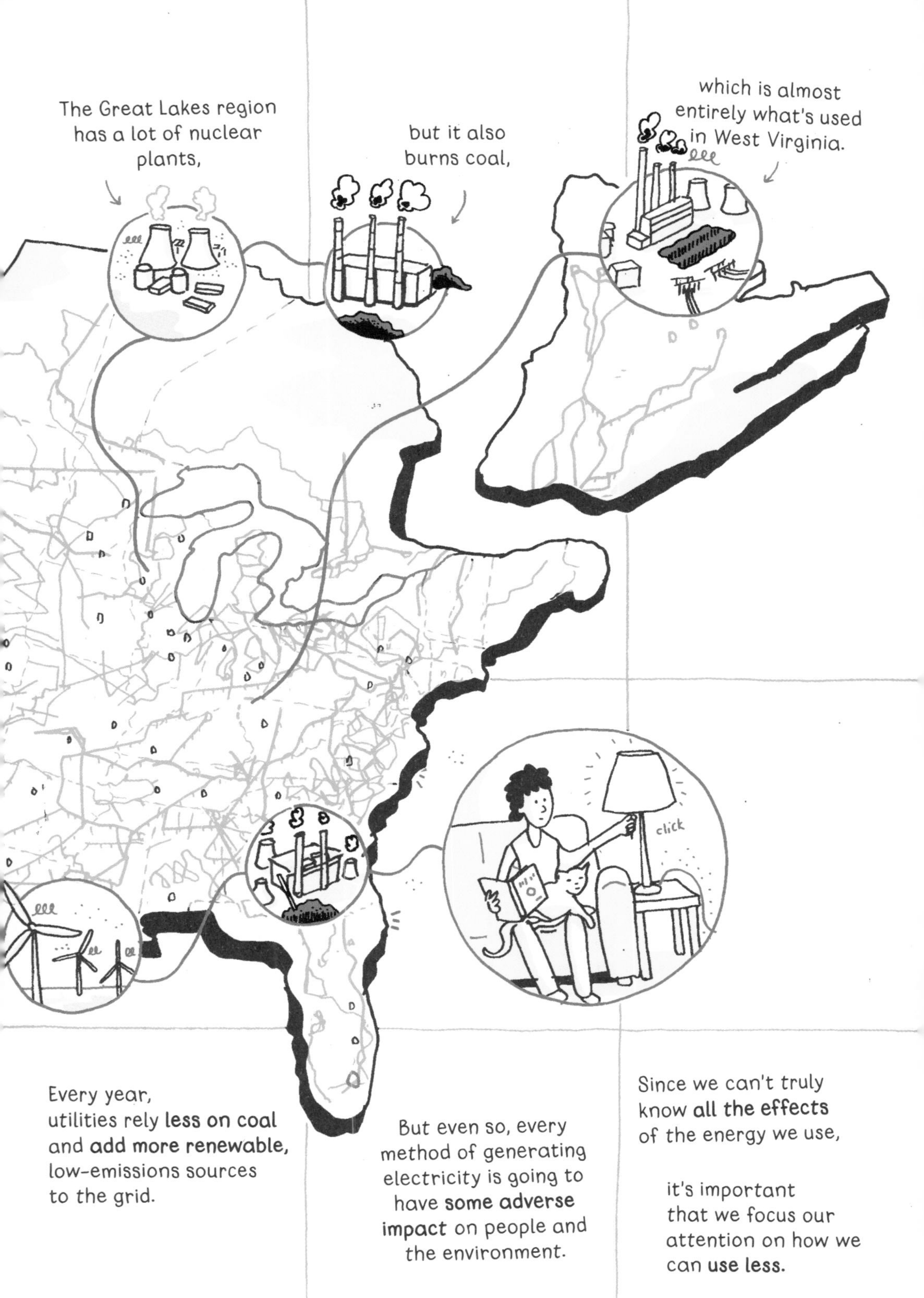
The Great Lakes region has a lot of nuclear plants,
but it also burns coal,
which is almost entirely what's used in West Virginia.
click
Every year, utilities rely less on coal and add more renewable, low-emissions sources to the grid.
But even so, every method of generating electricity is going to have some adverse impact on people and the environment.
Since we can't truly know all the effects of the energy we use,
it's important that we focus our attention on how we can use less.

Balancing the Grid: Distribution and Demand

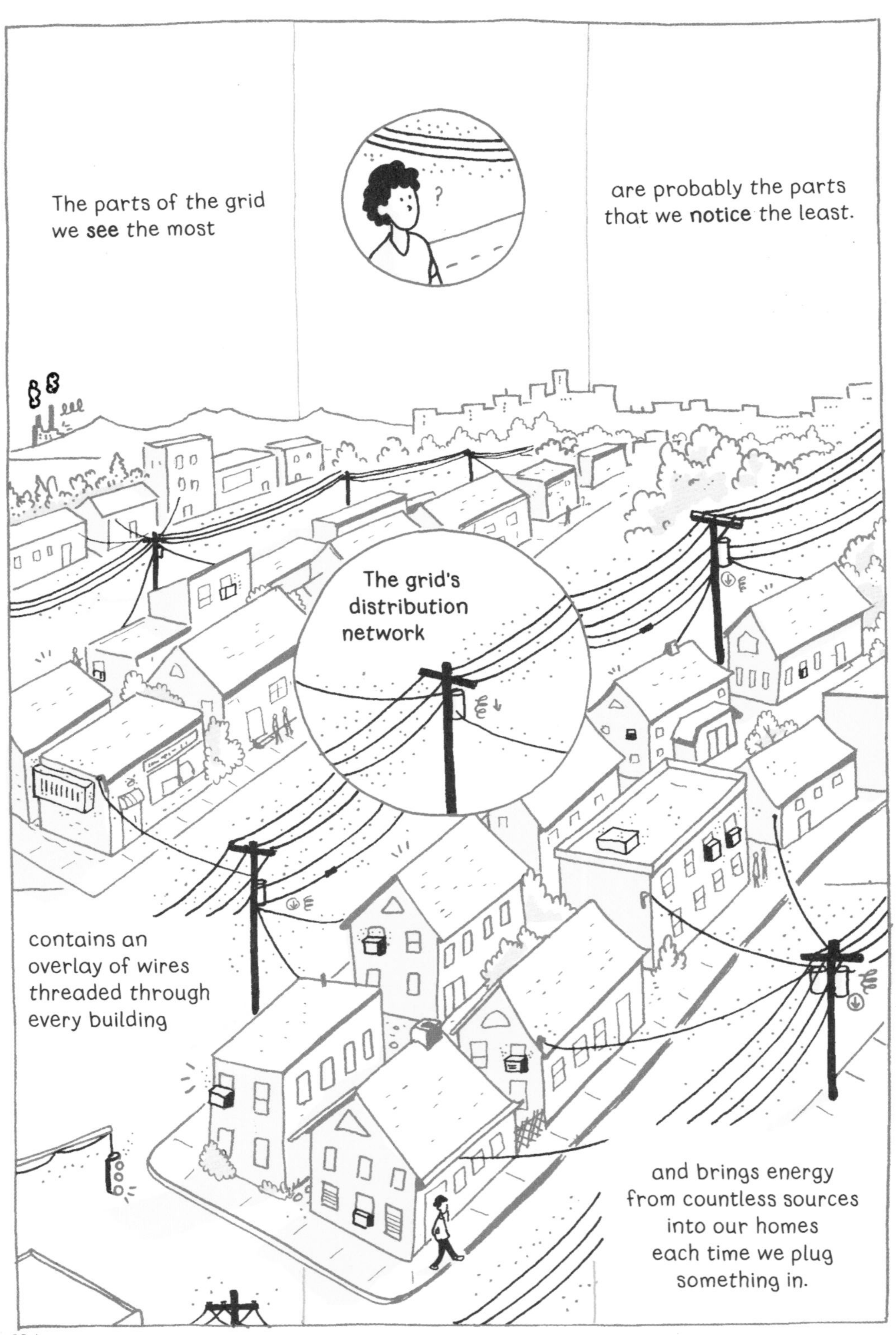
The parts of the grid
we **see** the most
?
are probably the parts
that we **notice** the least.
The grid's
distribution
network
contains an
overlay of wires
threaded through
every building
and brings energy
from countless sources
into our homes
each time we plug
something in.

In rural and suburban areas, electricity is usually propped up on **utility poles.**
primary high voltage wires for longer distance
the higher the voltage, the larger the insulators
high voltage
transformed to a lower voltage for homes, etc.
lower voltage
where the cable is spliced together
analog message board
lowest line is for phone and data
made from a tree that grows very straight
In urban areas, electricity lines are mostly **underground.**
Thomas Edison was right about this one.
New York City has come a long way since the **Pearl Street Station** network.
The city has over **250,000** manholes and more than **80,000** miles of cable,
providing an average of **8,000 MW** of electricity at any given time—and much more in the summer.

Our homes and electronics are also a part of this grid.
LED ~9W
When you plug in an air conditioner,
~1,000 watts (1 KW)
WRRRR
you're adding one more piece to a **vast network of wires and machines,** including:
the **outlet**
and the **wires** inside your home,
a **meter** outside,
$
so the utility knows how much $$ to charge
transformers,
step down
lowers voltage
local distribution lines,
medium voltage
lower voltage
substations,
switches, breakers, and transformers
high voltage

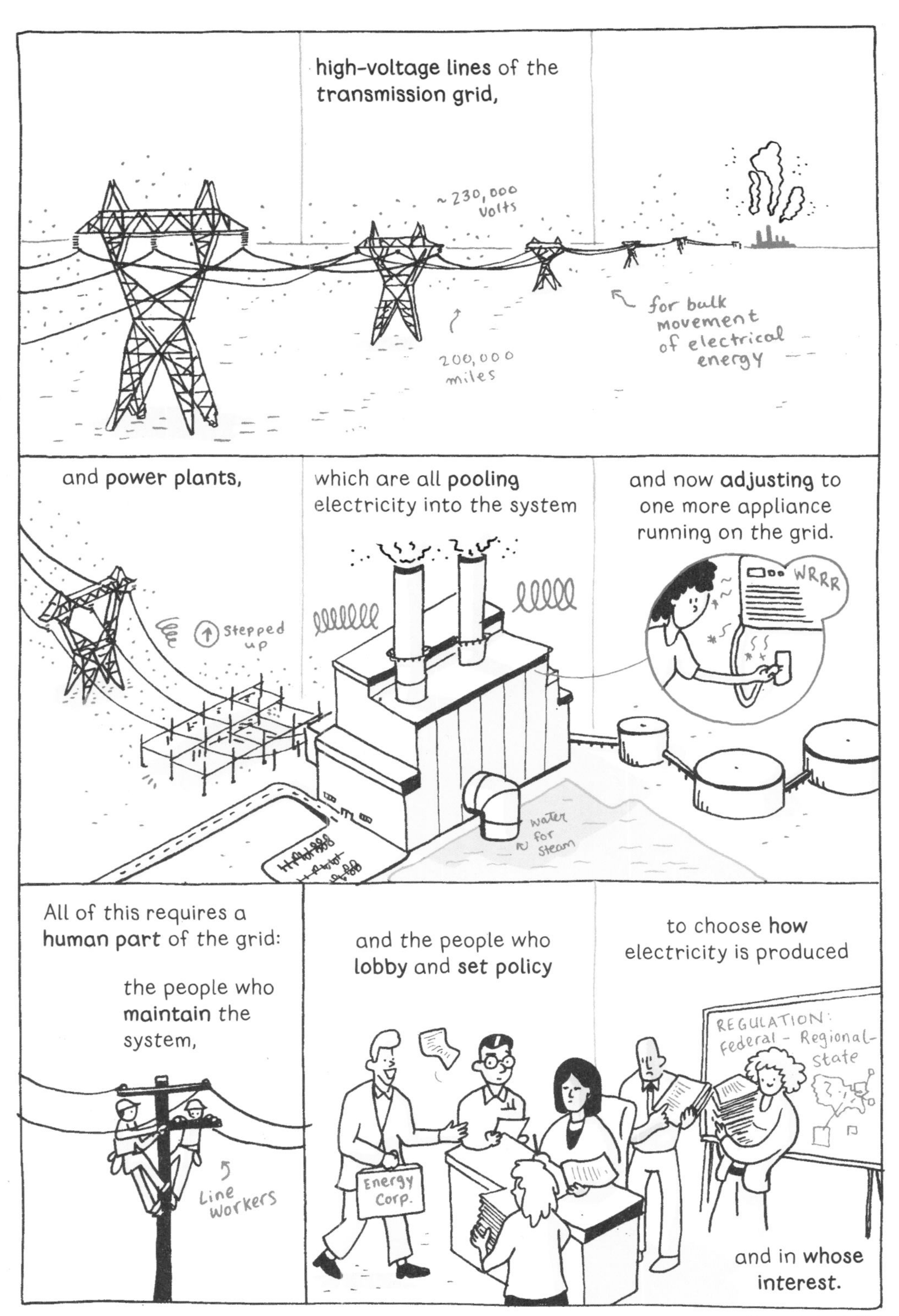
high-voltage lines of the transmission grid,
~230,000 volts
200,000 miles
for bulk movement of electrical energy
and power plants,
stepped up
which are all pooling electricity into the system
and now adjusting to one more appliance running on the grid.
WRRR
water for steam
All of this requires a human part of the grid:
the people who maintain the system,
Line workers
and the people who lobby and set policy
Energy Corp.
to choose how electricity is produced
REGULATION: Federal - Regional - state
and in whose interest.

Every hour of the day,
in operations centers around the country,
people coordinate power plants to generate an amount of power
perfectly matching the amount that's being used at that moment—
VRR
59.9 HZ
AREA
ACE
These groups are called balancing authorities
and can be single utilities that provide electricity for an area,
or regional organizations that run for-profit marketplaces for buying and selling electricity.

adjusting in real time
to every electrical device in every home, business, and factory
across an entire region.
click
Each area is set up and regulated differently by a web of state, national, and international regulations,
but each of these organizations has the same basic mandate:
to keep the grid balanced by matching the supply of energy with the demand.

People's electricity use depends a lot on weather, but follows a basic cycle:
increasing as people wake up,
lights
coffee
remaining steady during the day,
office stuff
factories
peaking in the evening when everyone is home,
TVs
appliances
and falling again as people go to sleep.
For balancing supply and demand on the grid,
+/-
utilities usually built systems to run on coal, nuclear, and gas
because operators can run them as needed,
in any season,
at any time of day.
Renewables are trickier to rely on
drought
because it's hard to know exactly when the sun will shine or the wind will blow,
tweet
and often, they don't provide power at all.
very variable

Giant batteries can help us store this energy **chemically**
and put it back in the grid when it's **most needed.**
Bzz
Bzz
Bzz
But batteries come with their own **environmental costs,**
cobalt
nickel
lithium
and even the largest can't store **enough to replace most power plants**—yet.
There are a lot of creative ways to make a "battery" by **storing energy.**
Like pumping water **uphill** when there's **extra** electricity
Pumped Storage
uses power to pump
top reservoir
and **releasing** it to spin a generator during **peak** demand.
lower reservoir
Or by connecting lots of the **smaller** batteries inside of people's **homes** and **electric cars.**
By allowing utilities to use these batteries to **store** and **dispatch** electricity to the grid as a **whole,**
we can create **virtual power plants** using pieces of the grid that would normally be dormant.

By combining the power grid with other systems, like the internet, we can reduce energy use—
and save electricity from ever having to be generated.
Like by hooking up appliances to run only when there's **extra electricity** on the grid,
smart dishwasher
woosha woosha woosha
and allowing utilities to turn them **down**—
especially on hot days when the **grid is overloaded.**
BRRRRRrrr..
OFF
But it raises questions like:
data centers use huge amounts of electricity.
Who makes these decisions?
And how much **control** of our energy use are people willing to give up?
brr....
It also means that utilities now have to protect the power grid against **cyber warfare** and **hackers,**
who can cause **blackouts** by turning **off** electricity generation,
transformers
or by **harnessing** it to fry essential pieces of equipment.

Some areas use **microgrids,**
small versions of the grid that can power a **community** or a **campus**
or an **individual house,**
and can **connect** or **disconnect** from the large grid as needed.
Microgrids can be essential where a reliable big grid **isn't available,**
but they shouldn't lead us to abandon solutions for the **public grid.**
We could risk returning to a time where only **some** had access,
19th-century mansion
Private Power
and miss an opportunity to **reimagine** and **rebuild** our grid
by **unifying a fragmented system** or relying on **local renewables**
so that clean, abundant energy
batteries
can be used by anyone at any time, across the continent.

Conclusion
Our grid began **simply**, as scientists tinkered with machines that harnessed a mysterious force.
In the hands of businesses, governments, and communities, these machines evolved into local systems
that spread across the country
and the entire planet
and now power every part of our lives.

Around the world, grids are made of a **transmission** system,
carrying electricity at high voltage from places of **abundance**
to places of **demand.**
Generators
convert various sources of energy to electric energy
fuel
GAS
(sometimes by harnessing the sun's **natural energy**),
and **local lines distribute** that energy
by bringing electricity into each home at **lower voltage,**
purr
which we use for pretty much everything.

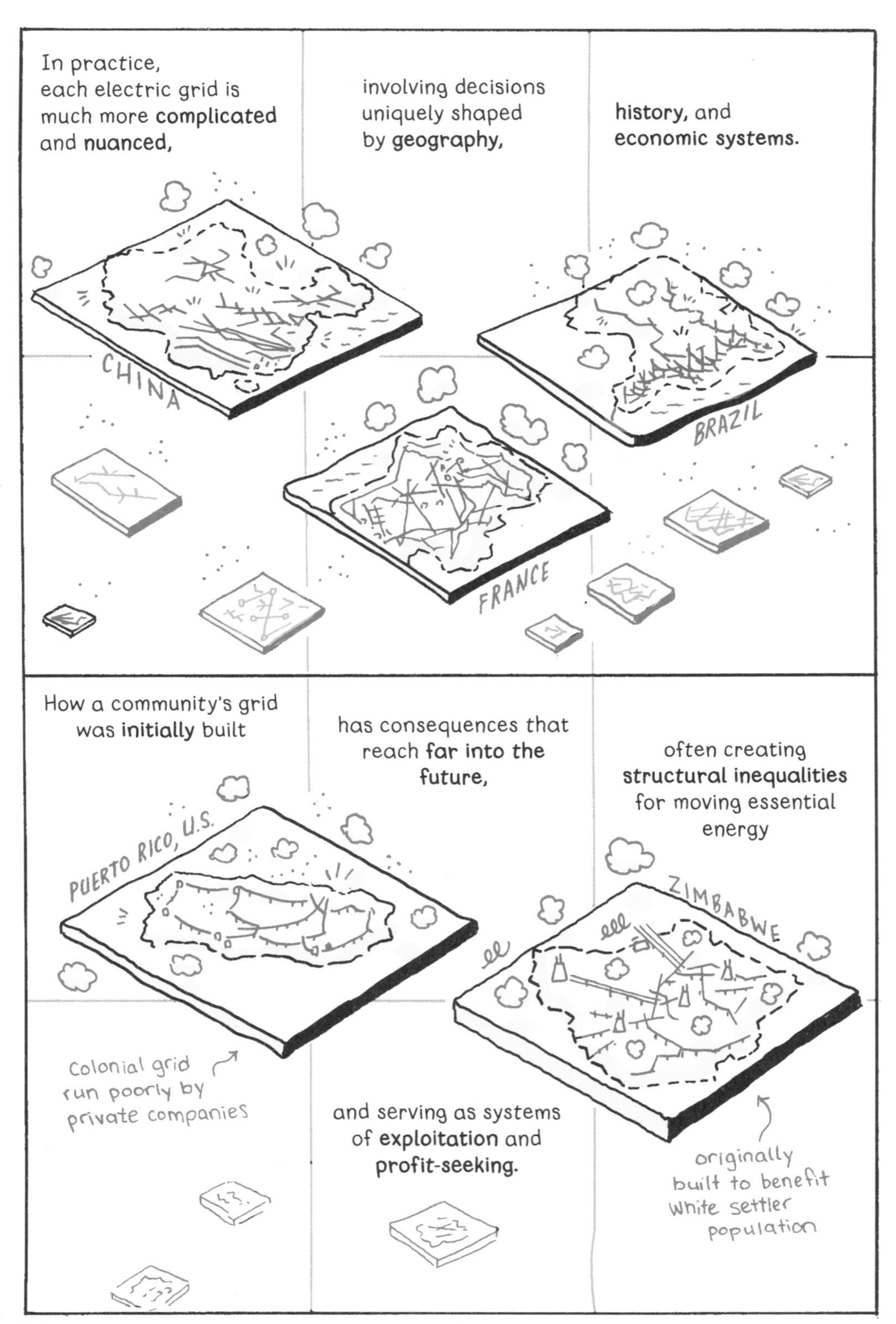
In practice,
each electric grid is
much more **complicated**
and **nuanced,**
involving decisions
uniquely shaped
by **geography,**
history, and
economic systems.
CHINA
FRANCE
BRAZIL
How a community's grid
was **initially** built
has consequences that
reach **far into the**
future,
often creating
structural inequalities
for moving essential
energy
PUERTO RICO, U.S.
ZIMBABWE
Colonial grid
run poorly by
private companies
and serving as systems
of **exploitation** and
profit-seeking.
originally
built to benefit
white settler
population

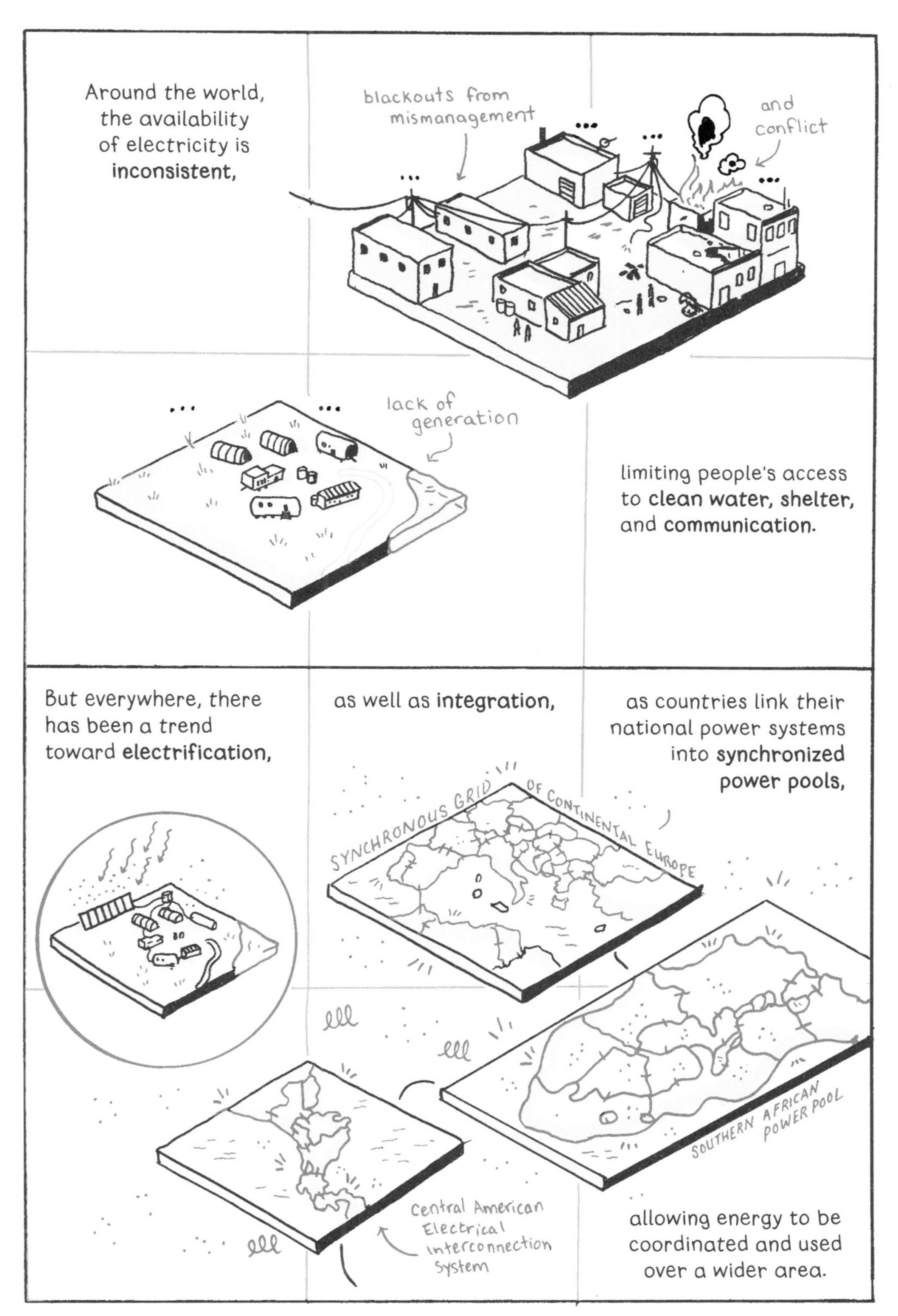
Around the world, the availability of electricity is **inconsistent,**
blackouts from mismanagement
and conflict
lack of generation
limiting people's access to **clean water, shelter,** and **communication.**
But everywhere, there has been a trend toward **electrification,**
as well as **integration,**
as countries link their national power systems into **synchronized power pools,**
SYNCHRONOUS GRID OF CONTINENTAL EUROPE
SOUTHERN AFRICAN POWER POOL
Central American Electrical Interconnection System
allowing energy to be coordinated and used over a wider area.

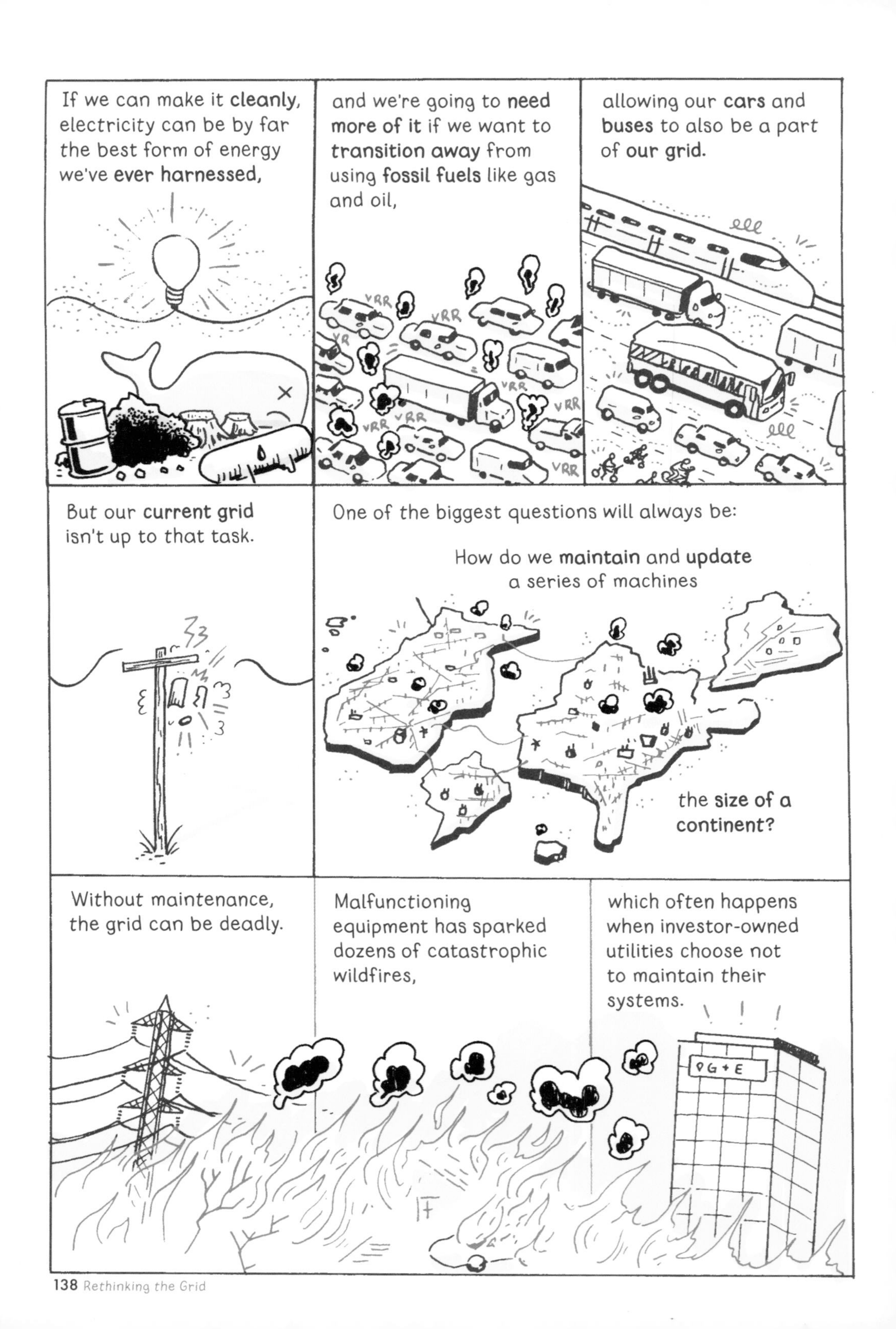
If we can make it cleanly, electricity can be by far the best form of energy we've ever harnessed,
and we're going to need more of it if we want to transition away from using fossil fuels like gas and oil,
VRR
VR
allowing our cars and buses to also be a part of our grid.
But our current grid isn't up to that task.
One of the biggest questions will always be:
How do we maintain and update a series of machines
the size of a continent?
Without maintenance, the grid can be deadly.
Malfunctioning equipment has sparked dozens of catastrophic wildfires,
which often happens when investor-owned utilities choose not to maintain their systems.
PG+E

An even bigger problem is the effect that generating electricity has on the planet.
Worldwide, generating electricity is the largest source of the greenhouse gases causing the Earth to warm.
In the U.S. and around the world, industries continue to invest in fossil fuels,
subjecting the world's most vulnerable people to its pollution
and to the most violent realities of a changing climate.
But these two crises—
our aging grid and the warming planet—
both require the same action.

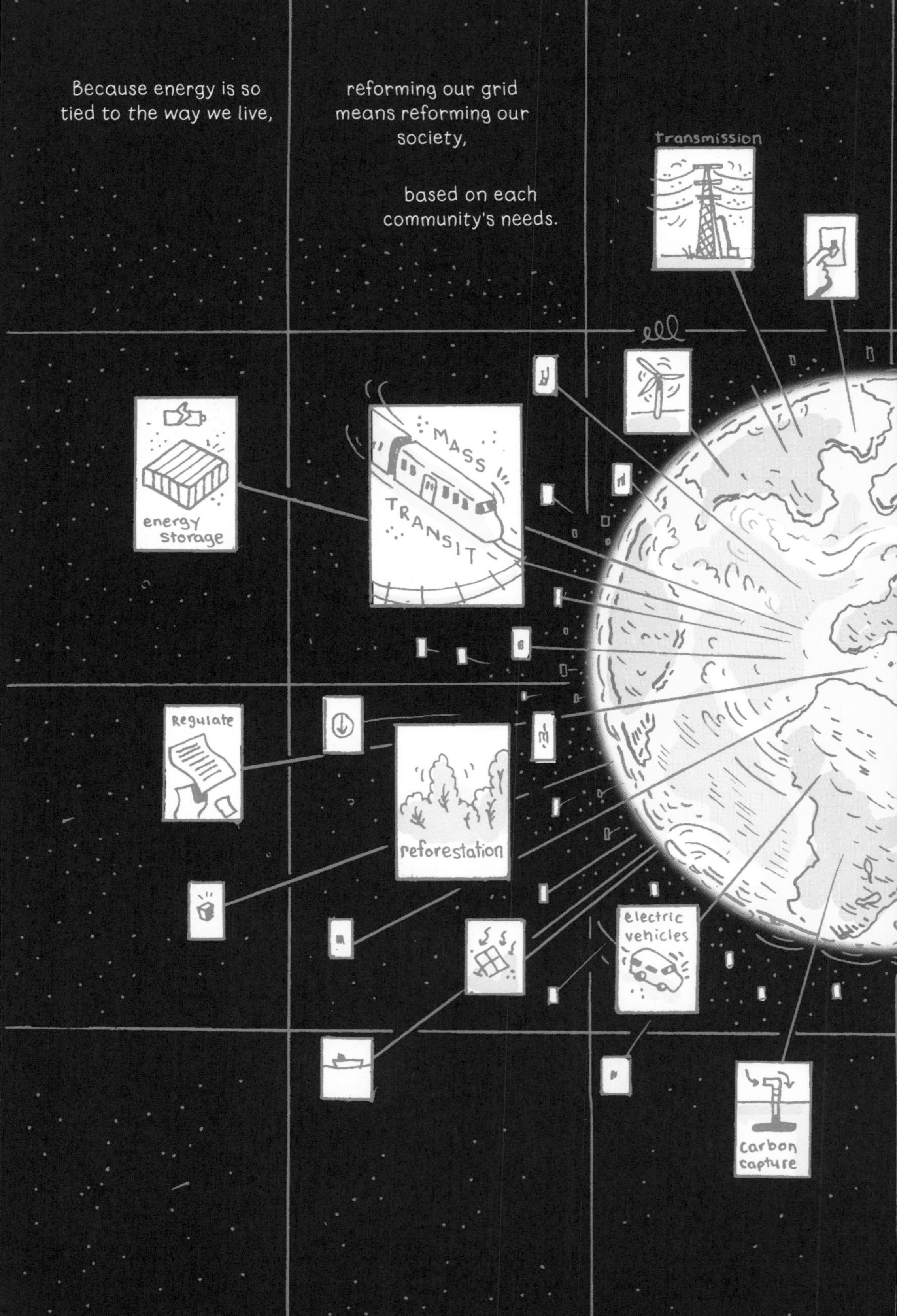
Because energy is so tied to the way we live,
reforming our grid means reforming our society,
based on each community's needs.
transmission
energy storage
MASS TRANSIT
Regulate
reforestation
electric vehicles
carbon capture

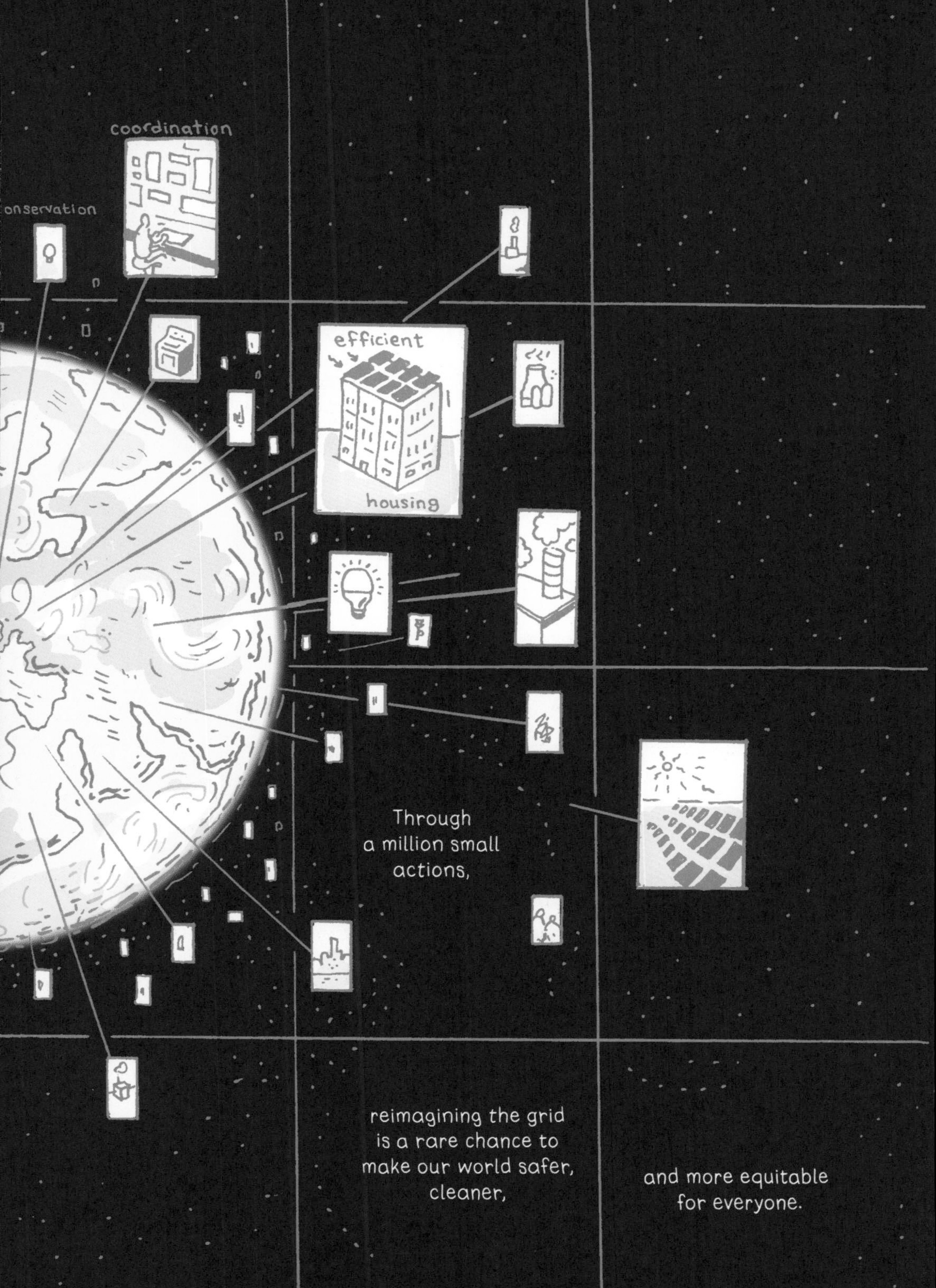
coordination
onservation
efficient
housing
Through a million small actions,
reimagining the grid is a rare chance to make our world safer, cleaner,
and more equitable for everyone.

Whoaaaa...

You know, they straightened out the Mississippi River in places, to make room for houses and livable acreage. Occasionally the river floods these places. "Floods" is the word they use, but in fact it is not flooding; it is remembering. Remembering where it used to be. All water has a perfect memory and is forever trying to get back to where it was.

—Toni Morrison

6 Waterworks

Where is our place in Earth's most important system?

SPLSH

pt
?

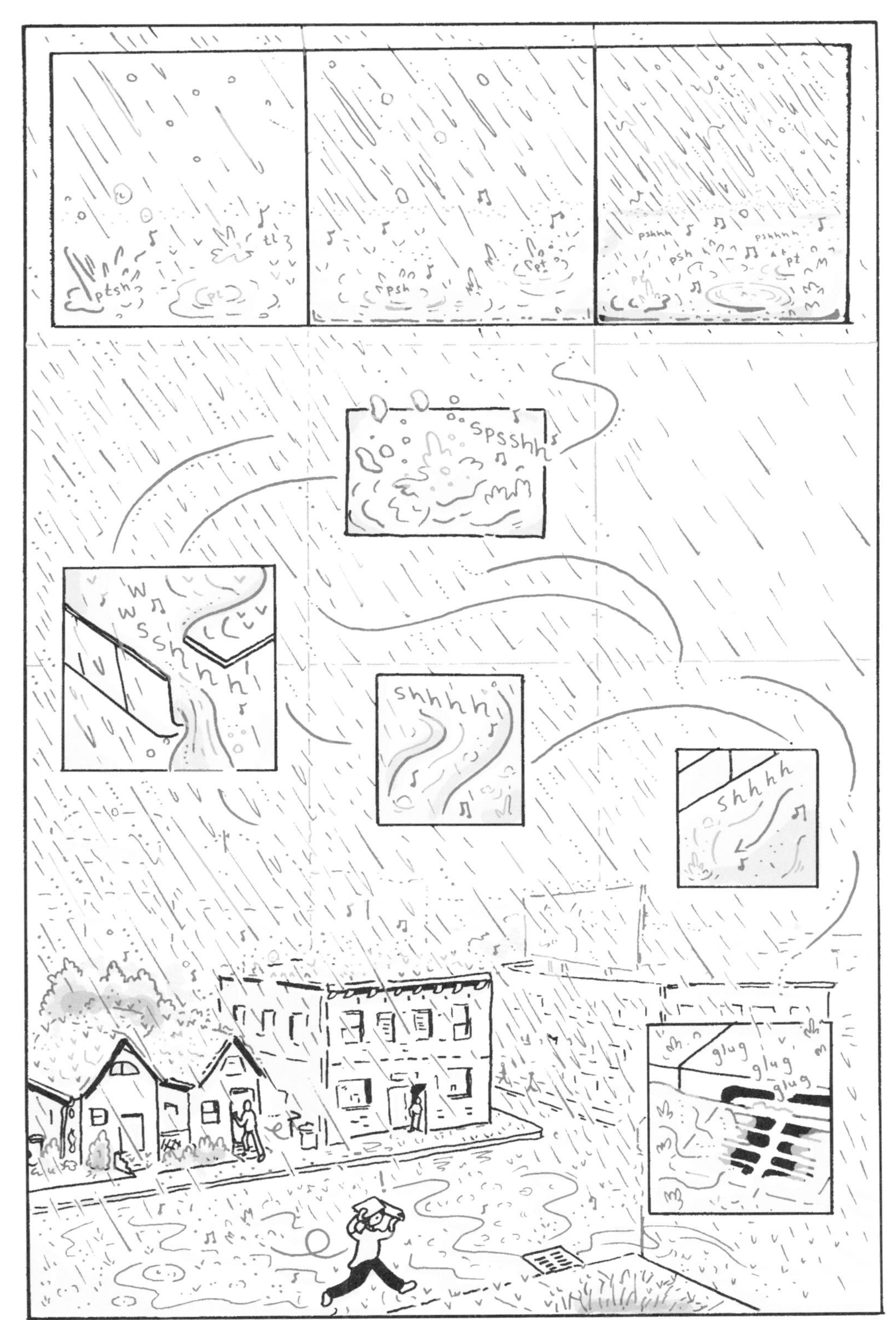
ptsh
pt
tl
psh
pt
pshhh
psh
pt
pshhhh
pt
spsshh
w
w
ssh
h
h
shhhh
shhhh
glug
glug
glug

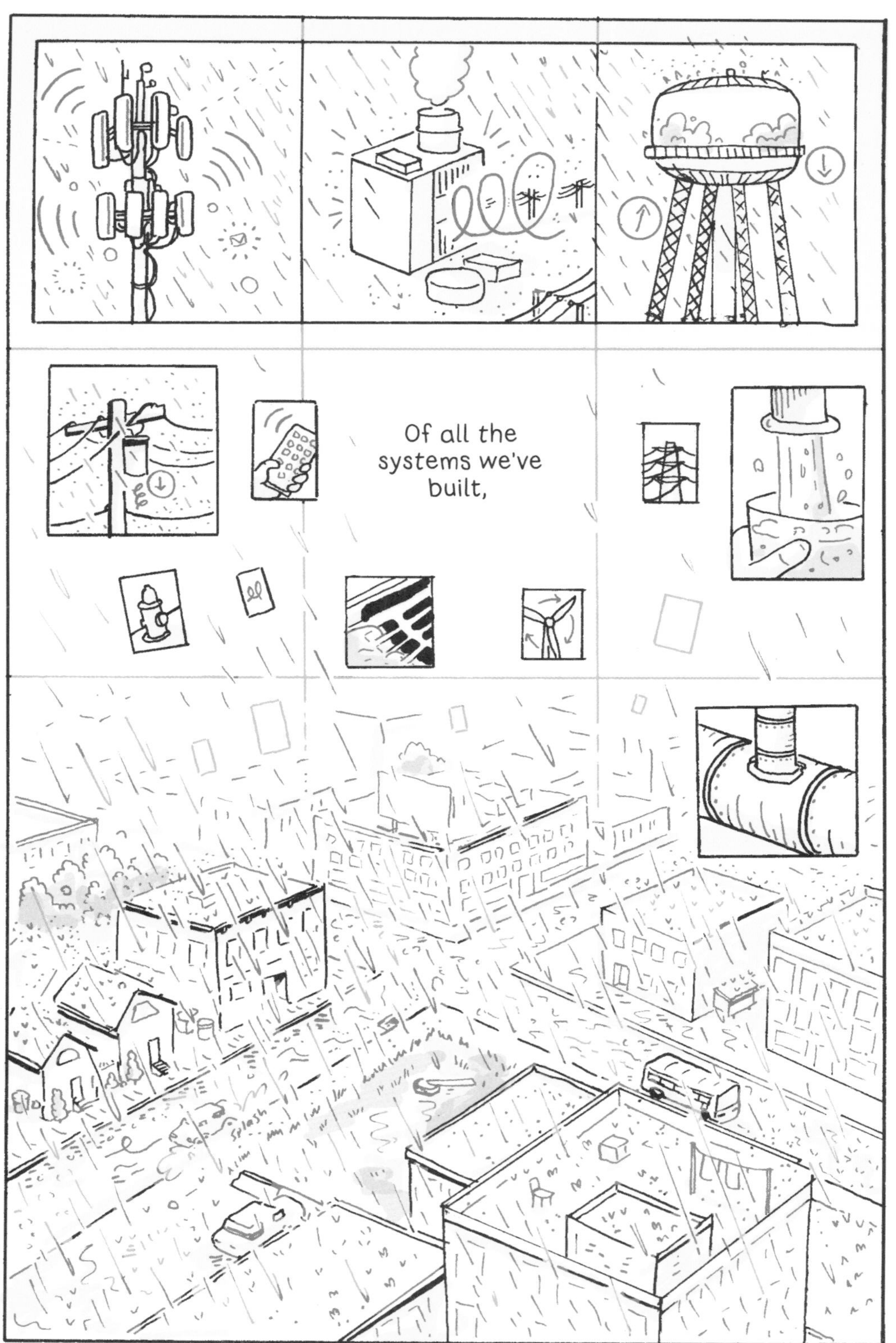
Of all the systems we've built,
Splash

none
compare to
the ones we
didn't.
WOOOOSH
CRACK

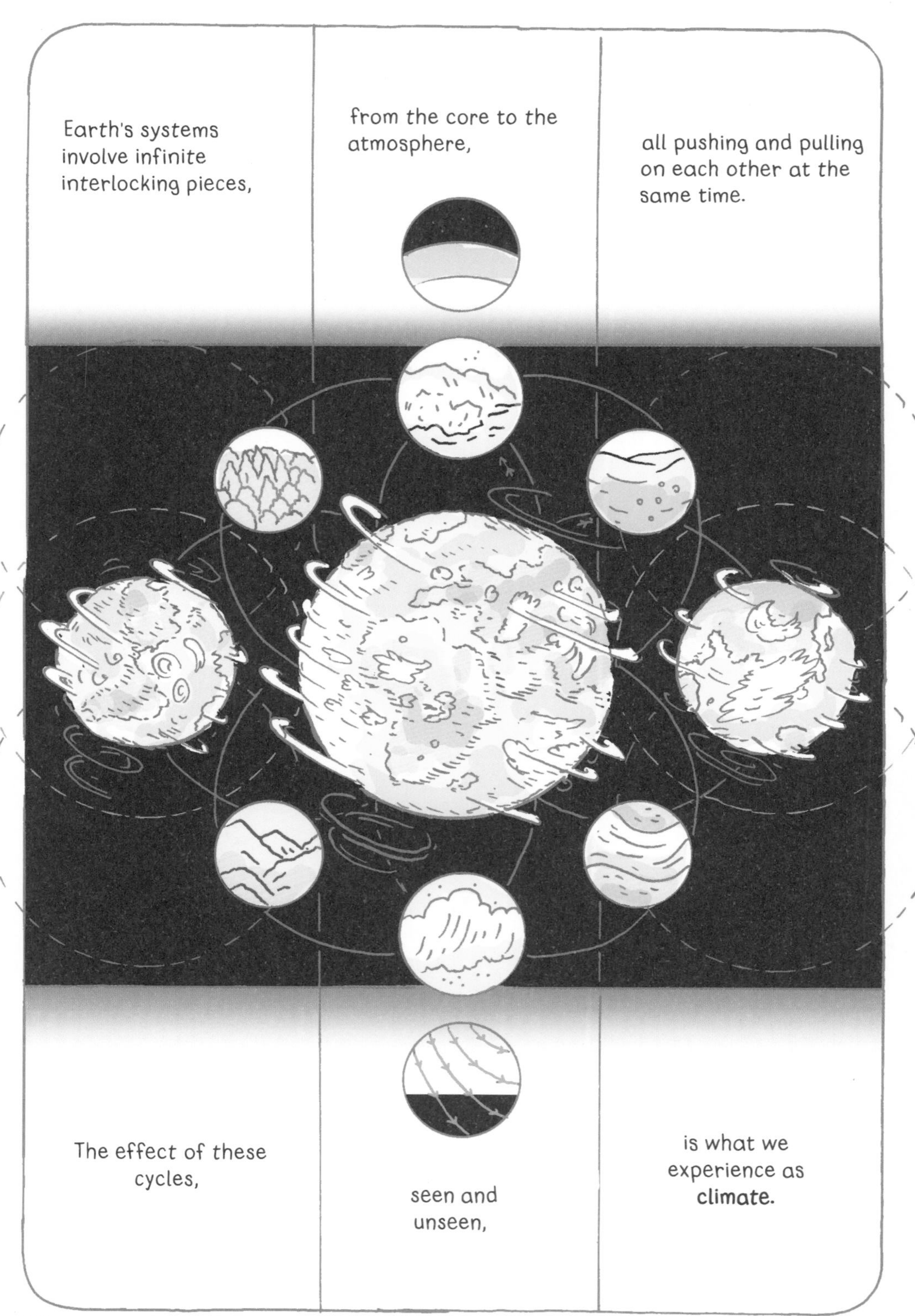
Earth's systems involve infinite interlocking pieces,
from the core to the atmosphere,
all pushing and pulling on each other at the same time.
The effect of these cycles,
seen and unseen,
is what we experience as **climate.**

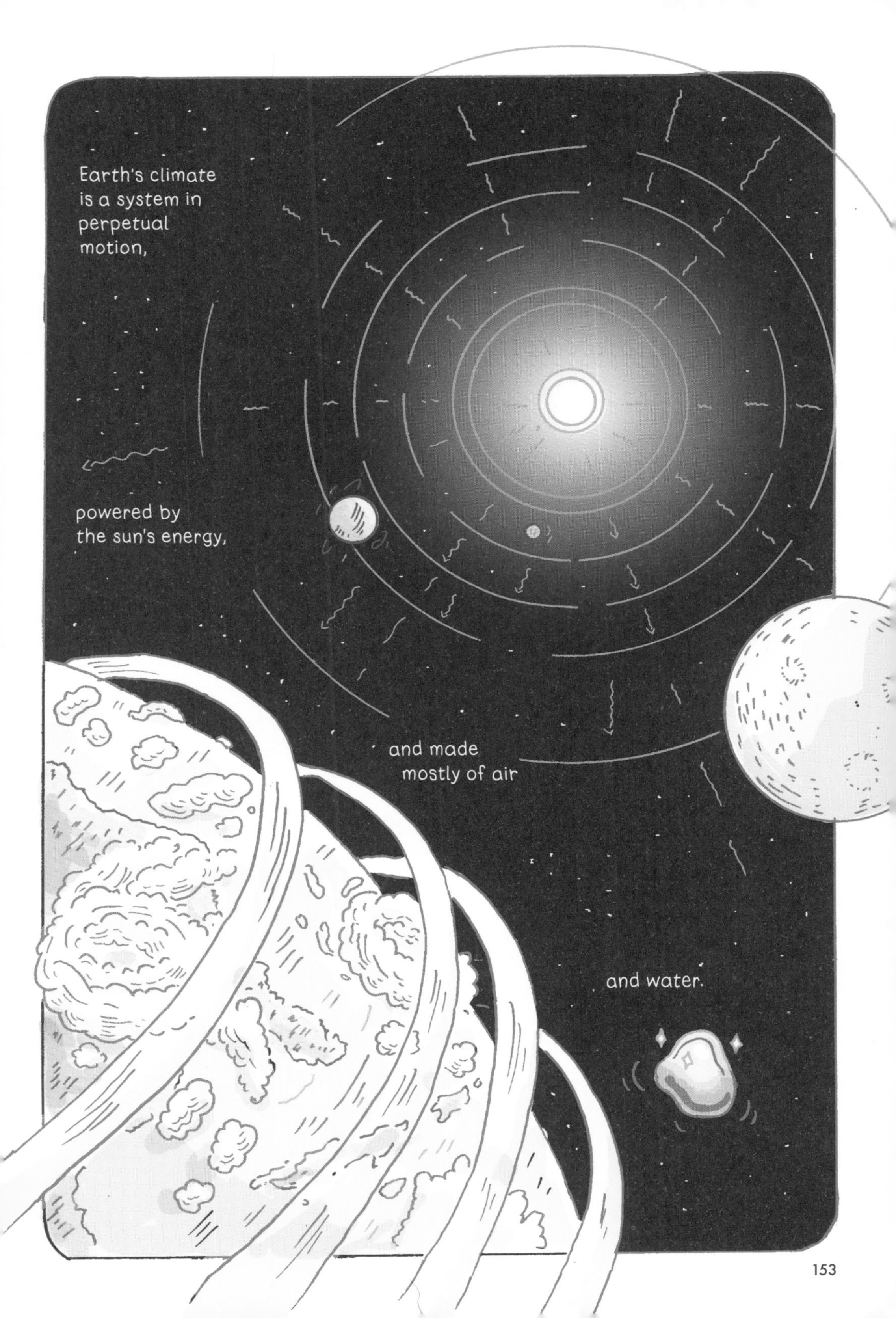
Earth's climate
is a system in
perpetual
motion,
powered by
the sun's energy,
and made
mostly of air
and water.

Water is,

sometimes,

for some of us,

the most boring thing we can think of

neutral taste

"fine"

not as good as seltzer

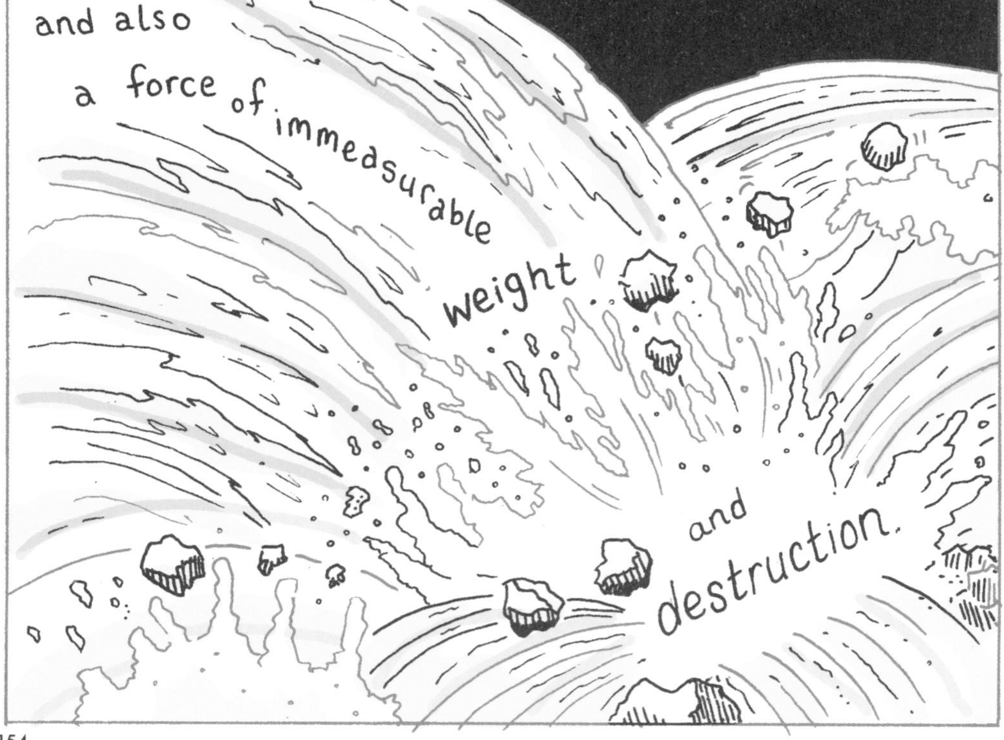

Water is a mysterious, spiritual, shape-shifting form of matter
at the root of all life
and a scarce resource used by every industry
to make products like paper and jeans and circuits,
to generate electricity,
and to carry waste somewhere else.

As the most **foundational** of all human needs,

there's beauty to the simplicity of **fresh, clean water,**

and people have **settled** where climate tended to bring them just the **right amount.**

But access to water has often required a **delicate balance**—

between **human systems**

waterworks

and **natural cycles.**

climate

Today, these two systems often provide water **unequally**—

both to different **geographies**

and to different **populations**—

and the **disparities** involved are growing wider.

The waterworks that underpin our lives—
for energy,
flushing,
cooking,
growing,
washing,
and drinking—
have always been built to hitch on to local climate cycles.
But as we make the Earth warmer and the climate more extreme,
each region now faces a higher risk of having too much water
or too little.
There are infinite ways to tell the story of our relationship with water—
but to understand a topic as massive as changing water cycles,
we need to seek out a much wider view.

1. Earth's Water: Cycles and Scales

Big Bang
~ 14 billion
years ago

While water is what defines our planet **today,**
it's not clear exactly **when—**
or **how—**
Earth got **all its water.**

Earth formed around **4.5 billion years ago,**
but the oceans may not have emerged for another billion years,
molten rock
and we're not sure where all that water came from.
Earth's water may have been brought by icy comets
no atmosphere yet
or emerged from the rocks
of our cooling planet.
"Rodinia"
~1 billion years ago
Over time,
"Snowball Earth"
~650 million years ago
Earth has gone through **radical transformations**
thriving ocean life
Cambrian Earth
~500 million years ago
defined by the state of its **water** and **climate.**
Jurassic Era
~170 million years ago
The **current balance** of water in its three states
~3 million years ago
sea level ~60 feet higher
2-3°C warmer
emerged relatively **recently**
~20,000 years ago
ice sheet
late ice age
and is what makes our Earth so livable today.
21st Century

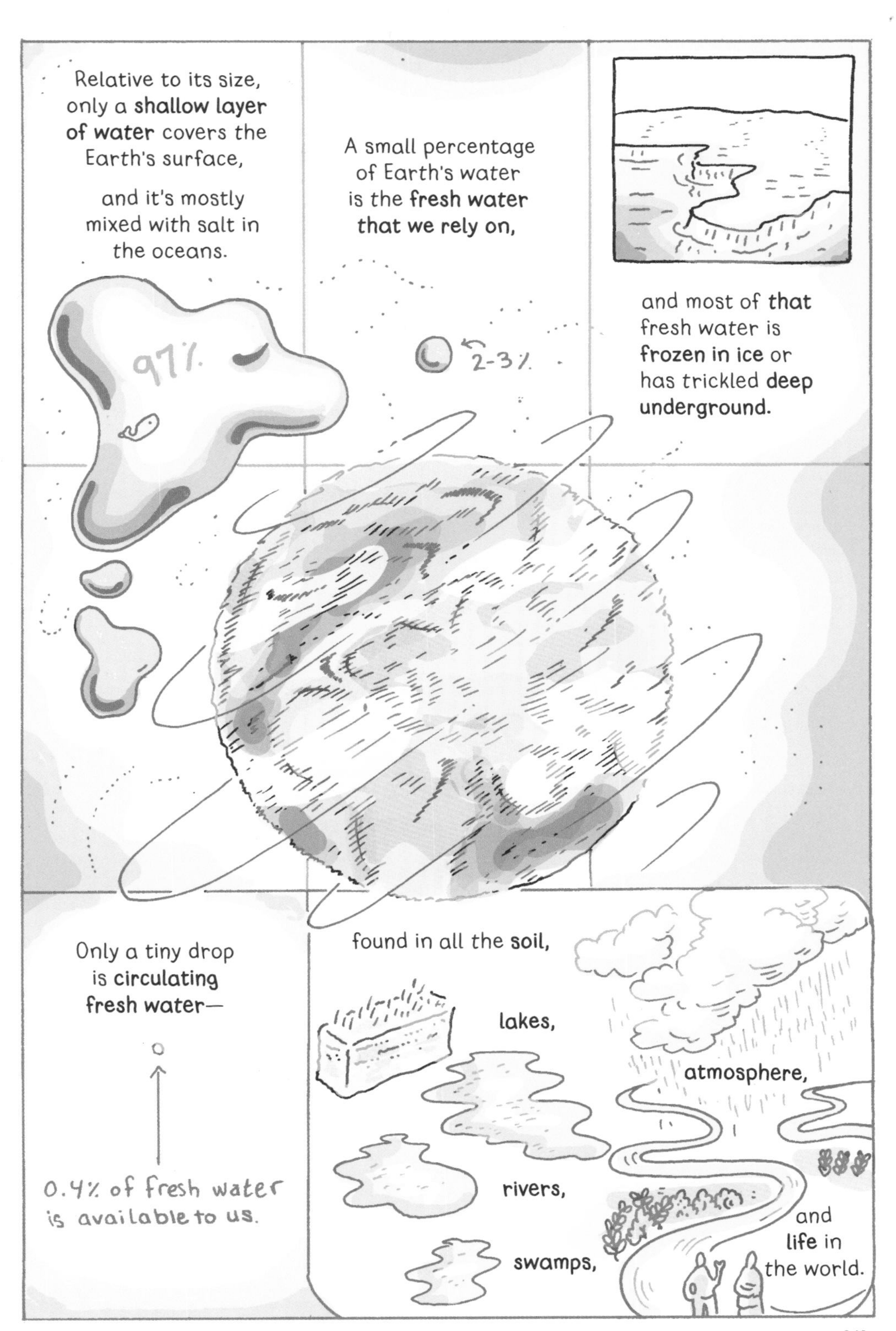
Relative to its size, only a shallow layer of water covers the Earth's surface,
and it's mostly mixed with salt in the oceans.
97%
A small percentage of Earth's water is the fresh water that we rely on,
2-3%
and most of that fresh water is frozen in ice or has trickled deep underground.
Only a tiny drop is circulating fresh water—
0.4% of fresh water is available to us.
found in all the soil,
lakes,
atmosphere,
rivers,
swamps,
and life in the world.

It's easy for us to **visualize** water on a **small scale**—

8 oz

to have a sense of the amount of **space** it takes up,

16 oz

~1 pound

and how much it **weighs.**

1 gallon

~8 pounds

This becomes difficult as we move away from our **typical momentary experiences** with water.

when did water get so heavy?

5 gallons
~40 pounds

And it's even harder to try and visualize the **amount we use in a day**

daily water use in U.S. per person

for basic chores like **flushing, bathing, washing dishes,** and **drinking.**

(~80-100 gallons)
~700 lbs

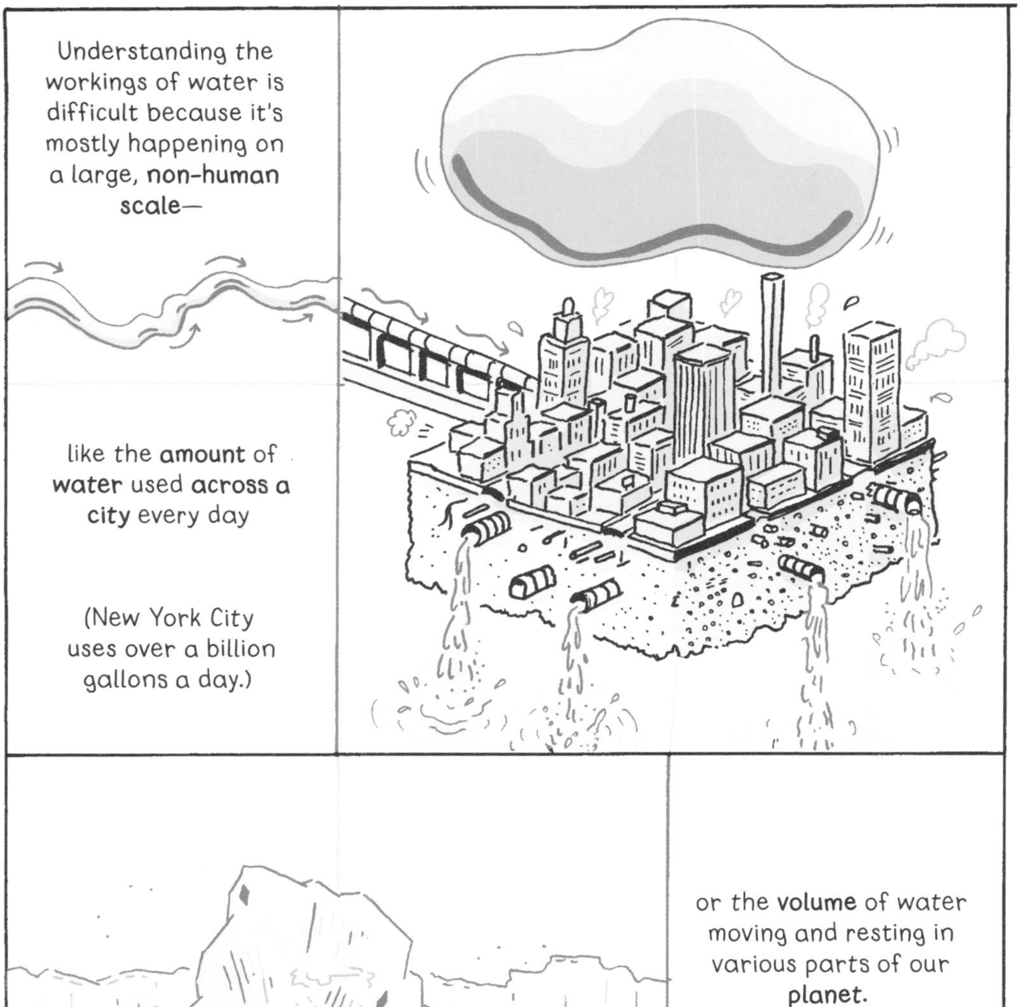

Understanding the workings of water is difficult because it's mostly happening on a large, **non-human scale**—

like the **amount** of **water** used **across a city** every day

(New York City uses over a billion gallons a day.)

or the **volume** of water moving and resting in various parts of our **planet.**

(There are 5.7 million "cubic miles" of fresh water frozen in glaciers and ice caps.)

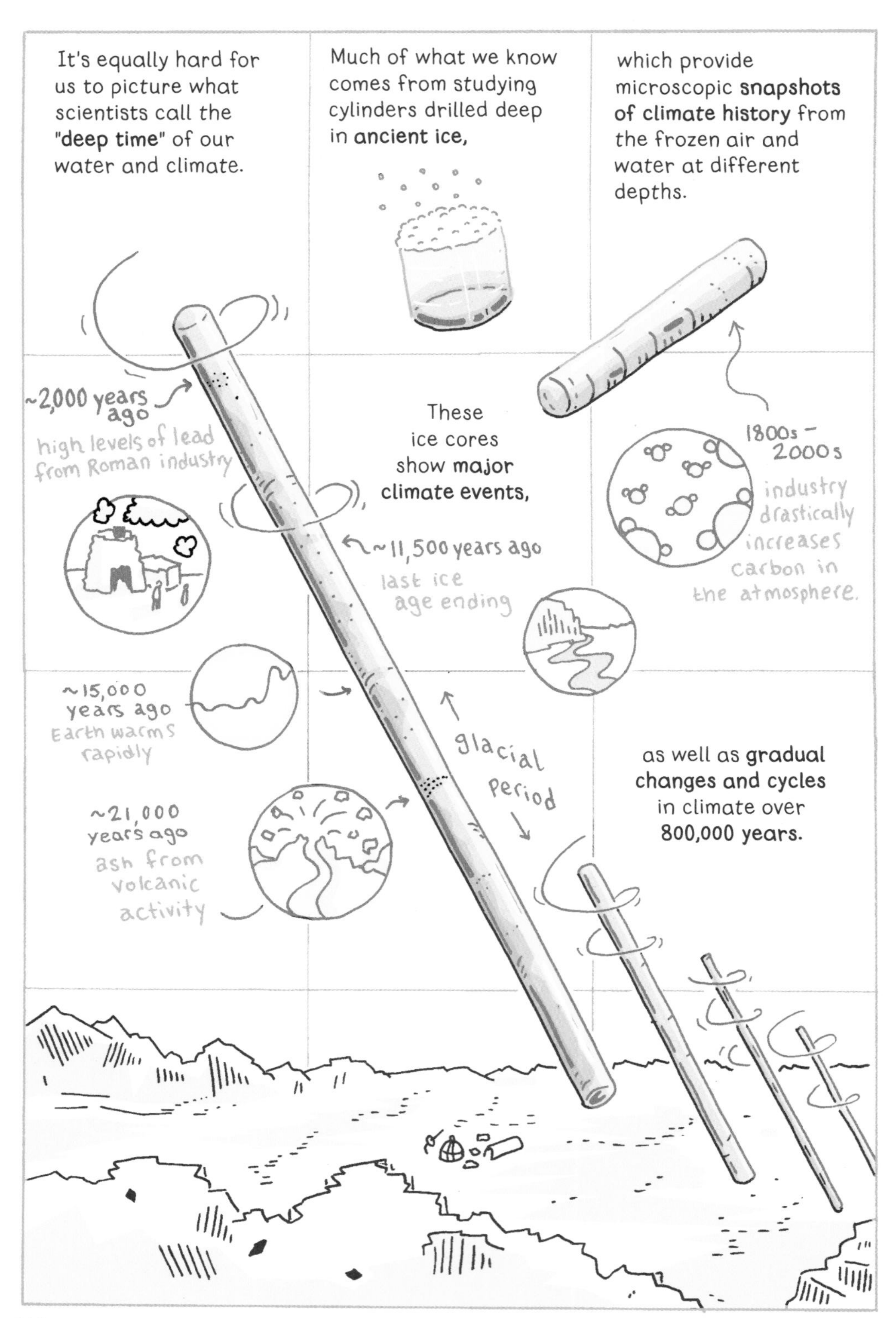
It's equally hard for us to picture what scientists call the "deep time" of our water and climate.
Much of what we know comes from studying cylinders drilled deep in ancient ice,
which provide microscopic snapshots of climate history from the frozen air and water at different depths.
~2,000 years ago
high levels of lead from Roman industry
These ice cores show major climate events,
1800s - 2000s
industry drastically increases carbon in the atmosphere.
~11,500 years ago
last ice age ending
~15,000 years ago
Earth warms rapidly
glacial period
as well as gradual changes and cycles in climate over 800,000 years.
~21,000 years ago
ash from volcanic activity

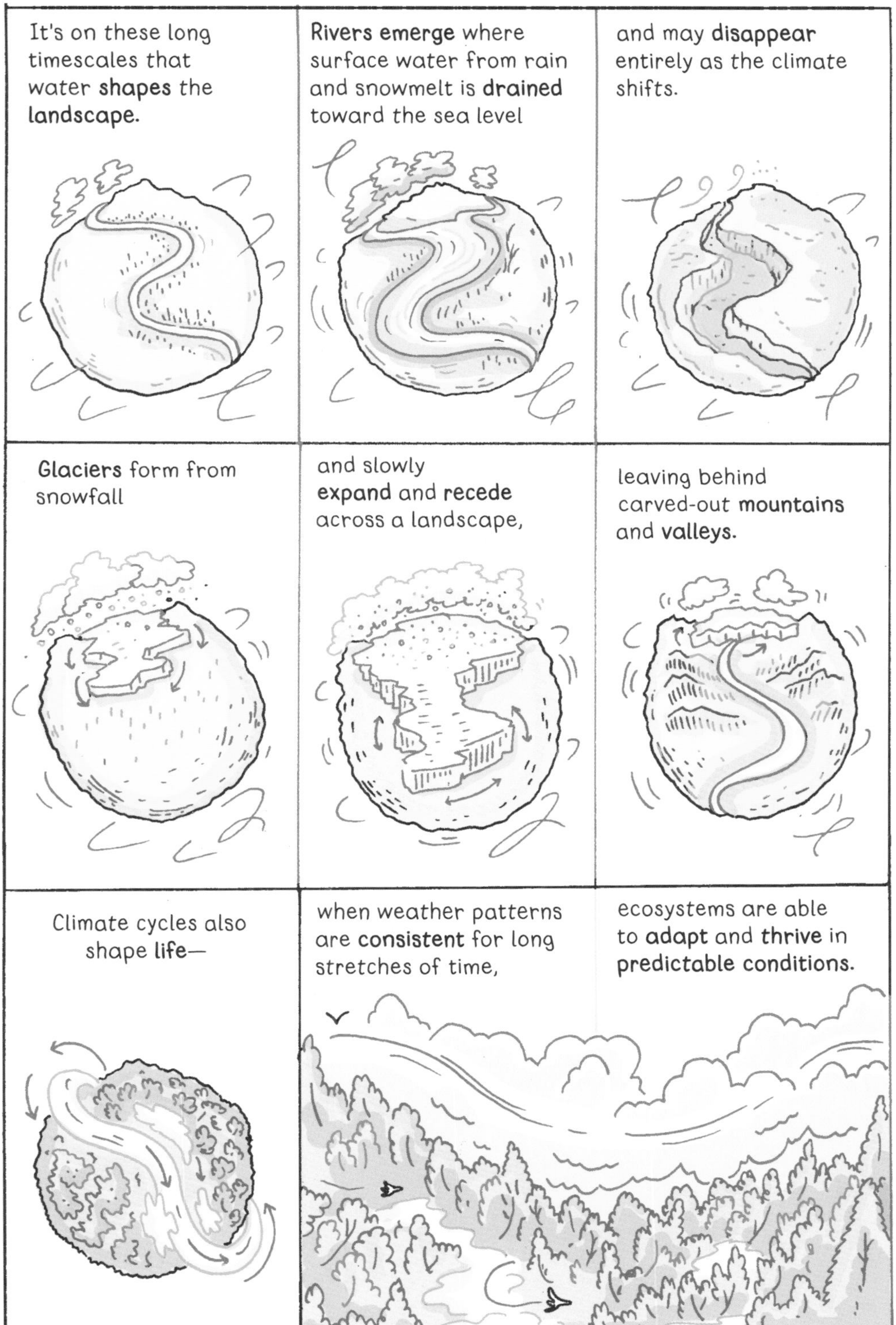

It's on these long timescales that water **shapes** the **landscape.**

Rivers emerge where surface water from rain and snowmelt is **drained** toward the sea level

and may **disappear** entirely as the climate shifts.

Glaciers form from snowfall

and slowly **expand** and **recede** across a landscape,

leaving behind carved-out **mountains** and **valleys.**

Climate cycles also shape **life**—

when weather patterns are **consistent** for long stretches of time,

ecosystems are able to **adapt** and **thrive** in **predictable conditions.**

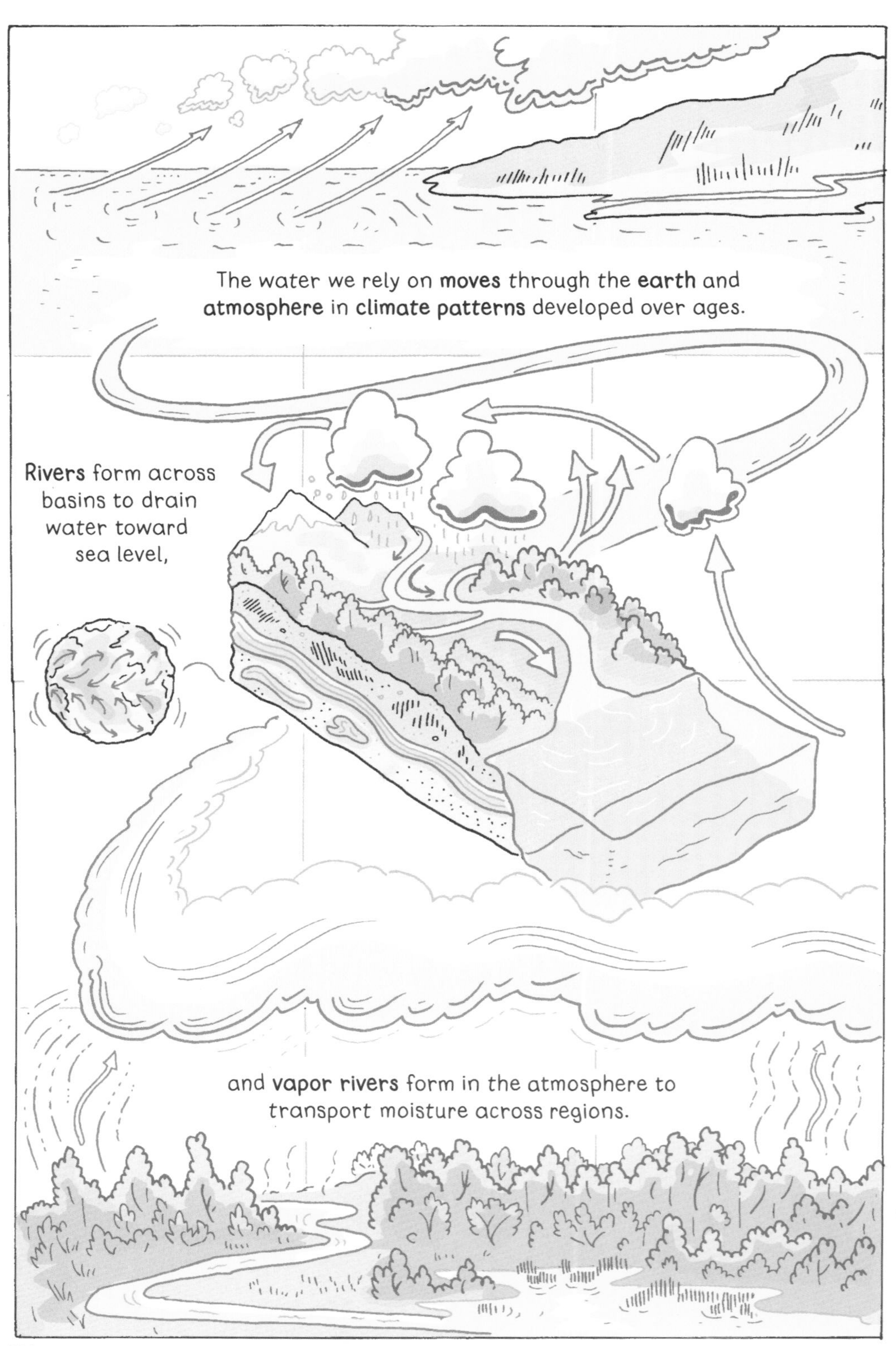

The water we rely on **moves** through the **earth** and **atmosphere** in **climate patterns** developed over ages.
Rivers form across basins to drain water toward sea level,
and **vapor rivers** form in the atmosphere to transport moisture across regions.

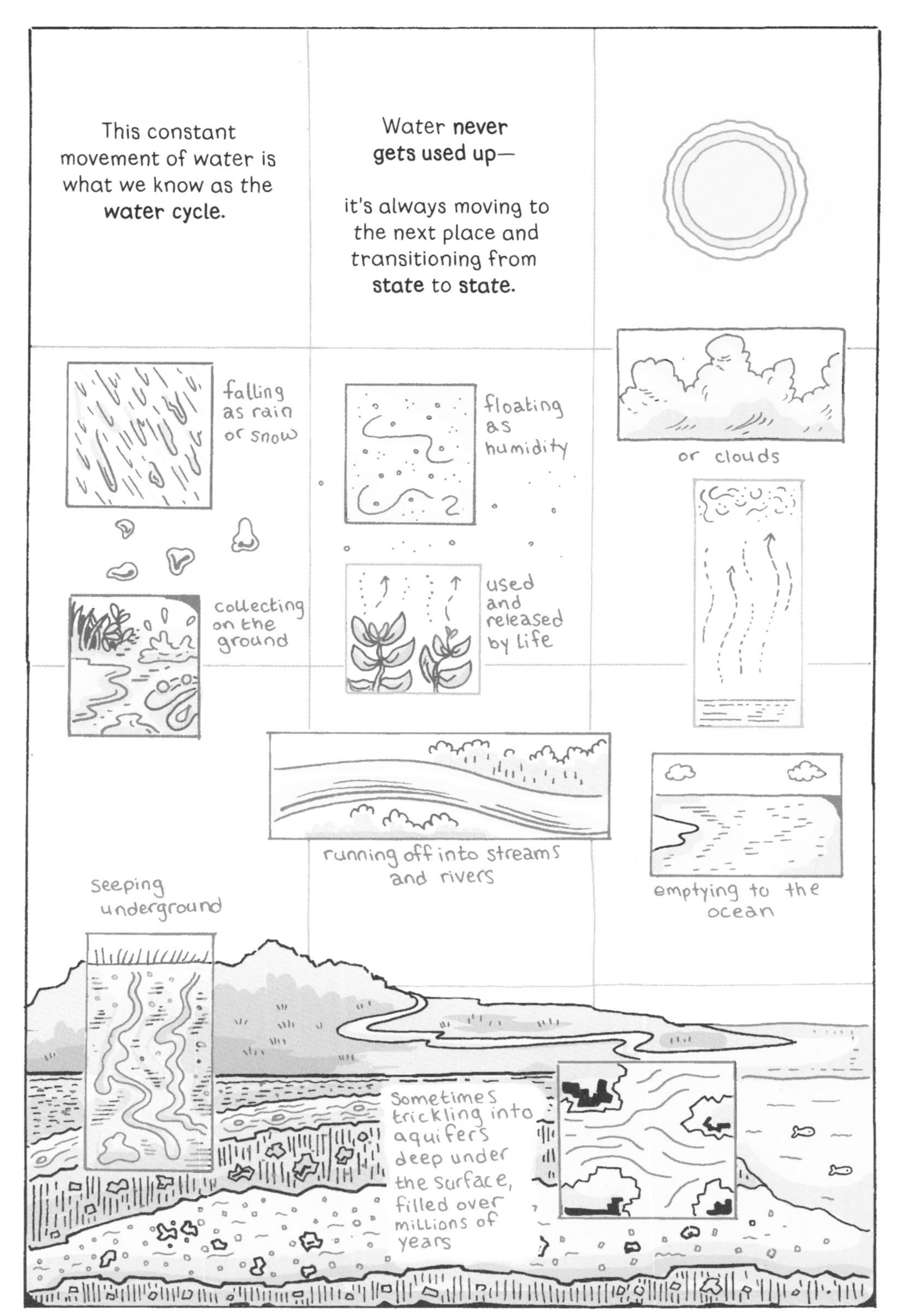
This constant movement of water is what we know as the **water cycle.**
Water **never gets used up**—
it's always moving to the next place and transitioning from **state** to **state.**
falling as rain or snow
floating as humidity
or clouds
collecting on the ground
used and released by life
running off into streams and rivers
emptying to the ocean
seeping underground
Sometimes trickling into aquifers deep under the surface, filled over millions of years

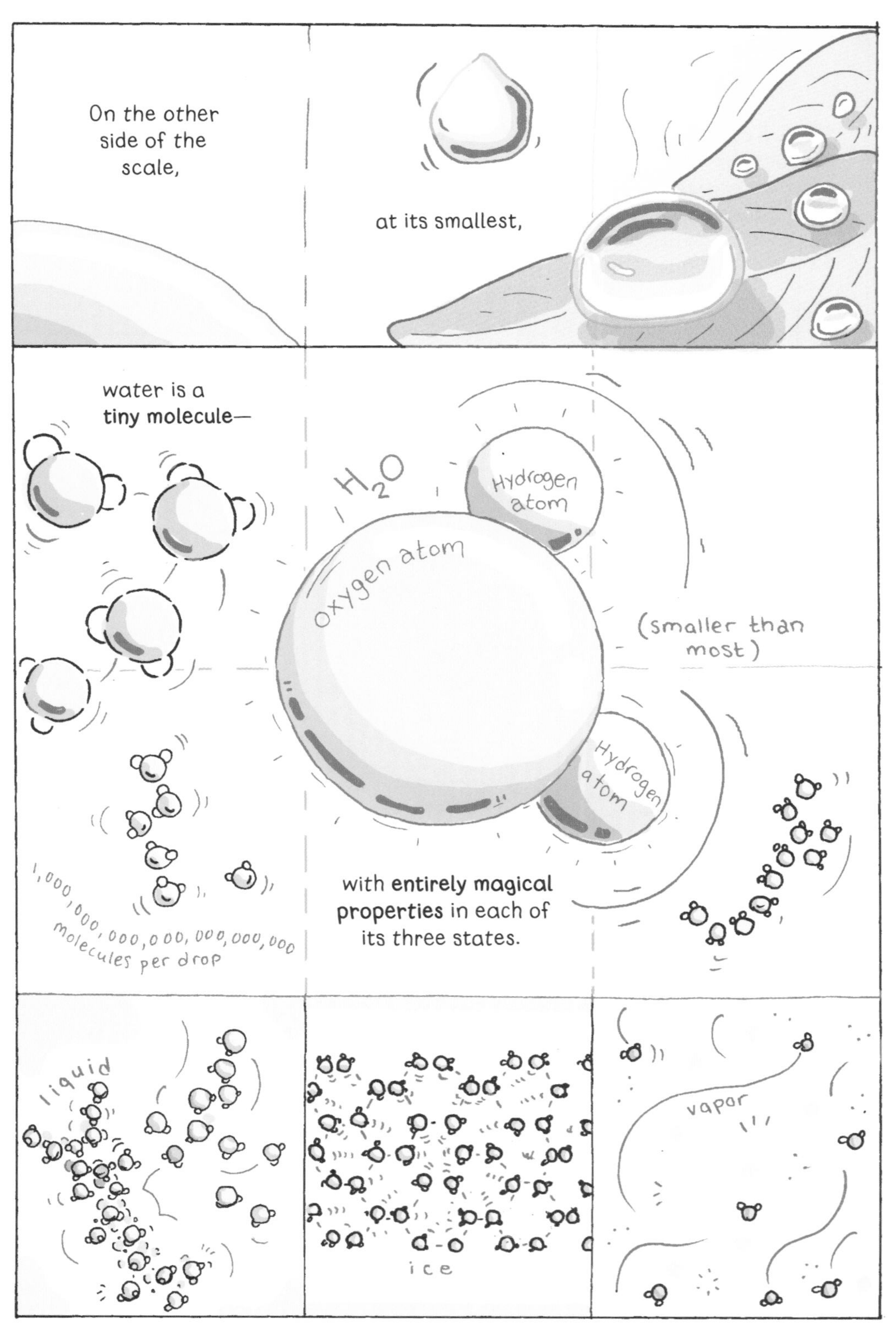
On the other side of the scale,
at its smallest,
water is a **tiny molecule**—
H_2O
Hydrogen atom
oxygen atom
(smaller than most)
Hydrogen atom
1,000,000,000,000,000,000,000 molecules per drop
with **entirely magical properties** in each of its three states.
liquid
ice
vapor

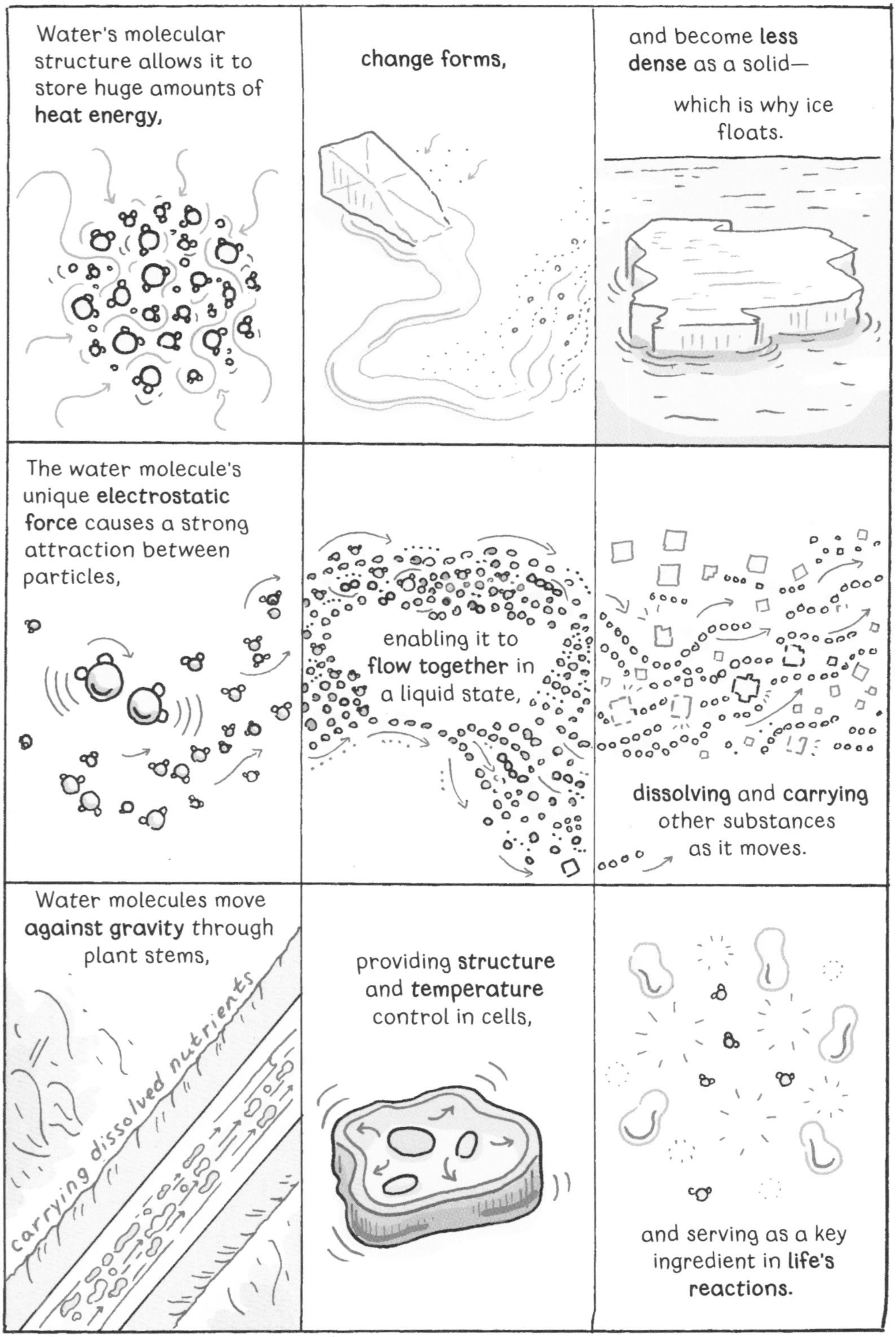

Water's molecular structure allows it to store huge amounts of **heat energy,**

change forms,

and become **less dense** as a solid—

which is why ice floats.

The water molecule's unique **electrostatic force** causes a strong attraction between particles,

enabling it to **flow together** in a liquid state,

dissolving and **carrying** other substances as it moves.

Water molecules move **against gravity** through plant stems,

carrying dissolved nutrients

providing **structure** and **temperature** control in cells,

and serving as a key ingredient in **life's reactions.**

Water's unique properties help it **regulate** the **temperature** of the **planet**
ice reflects sunlight
water absorbs it
and **transform** the land it falls on,

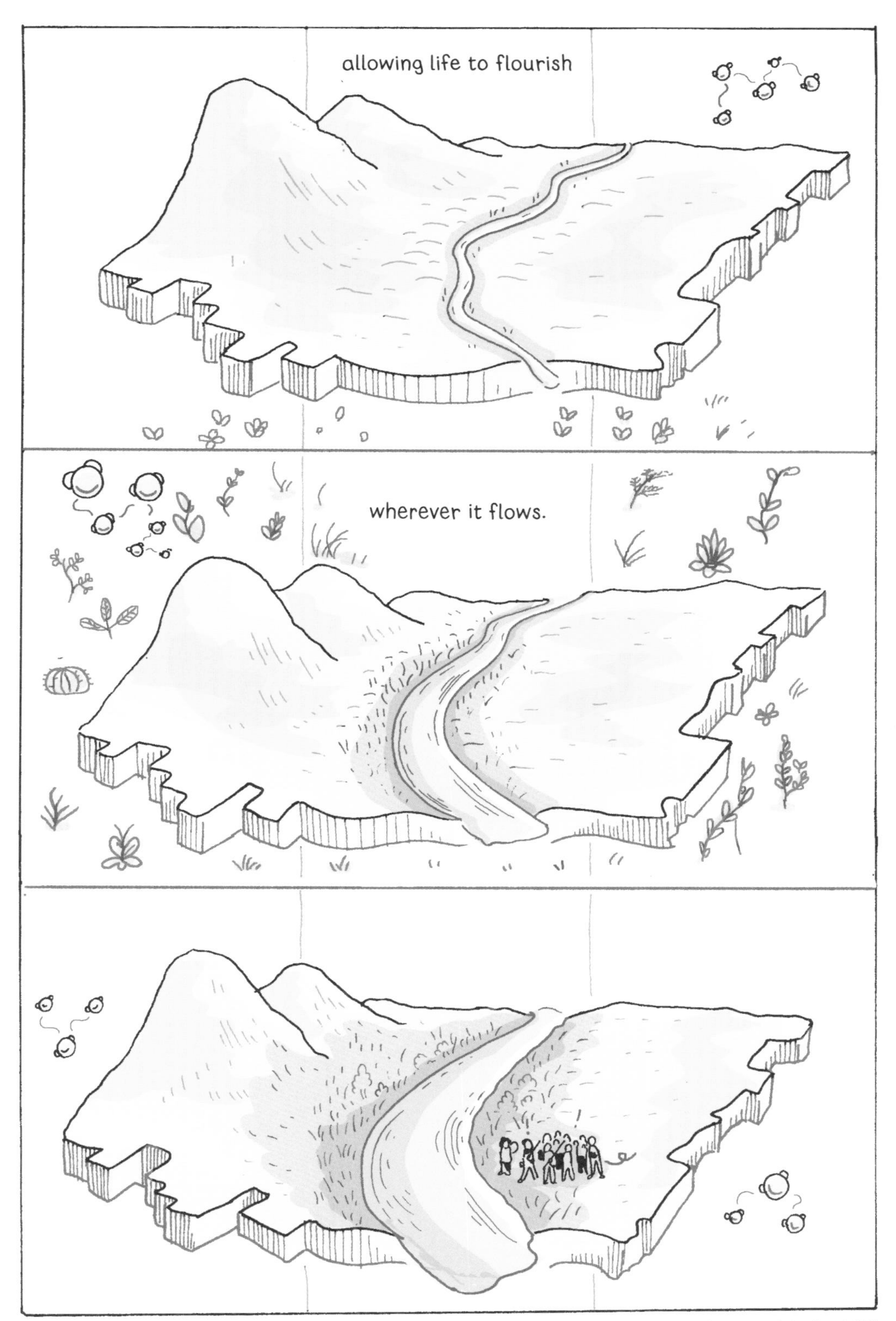
allowing life to flourish
wherever it flows.

On a **human**-sized scale,
water plays a crucial role in **our own** systems—
most importantly, in our bodies.
mostly water
Water from **eating** and **drinking** makes up **two-thirds** of our body's weight,
Crunch
and is **used** by cells, tissues, and organs
before being **released** through **breathing, sweating,** and **digestion.**

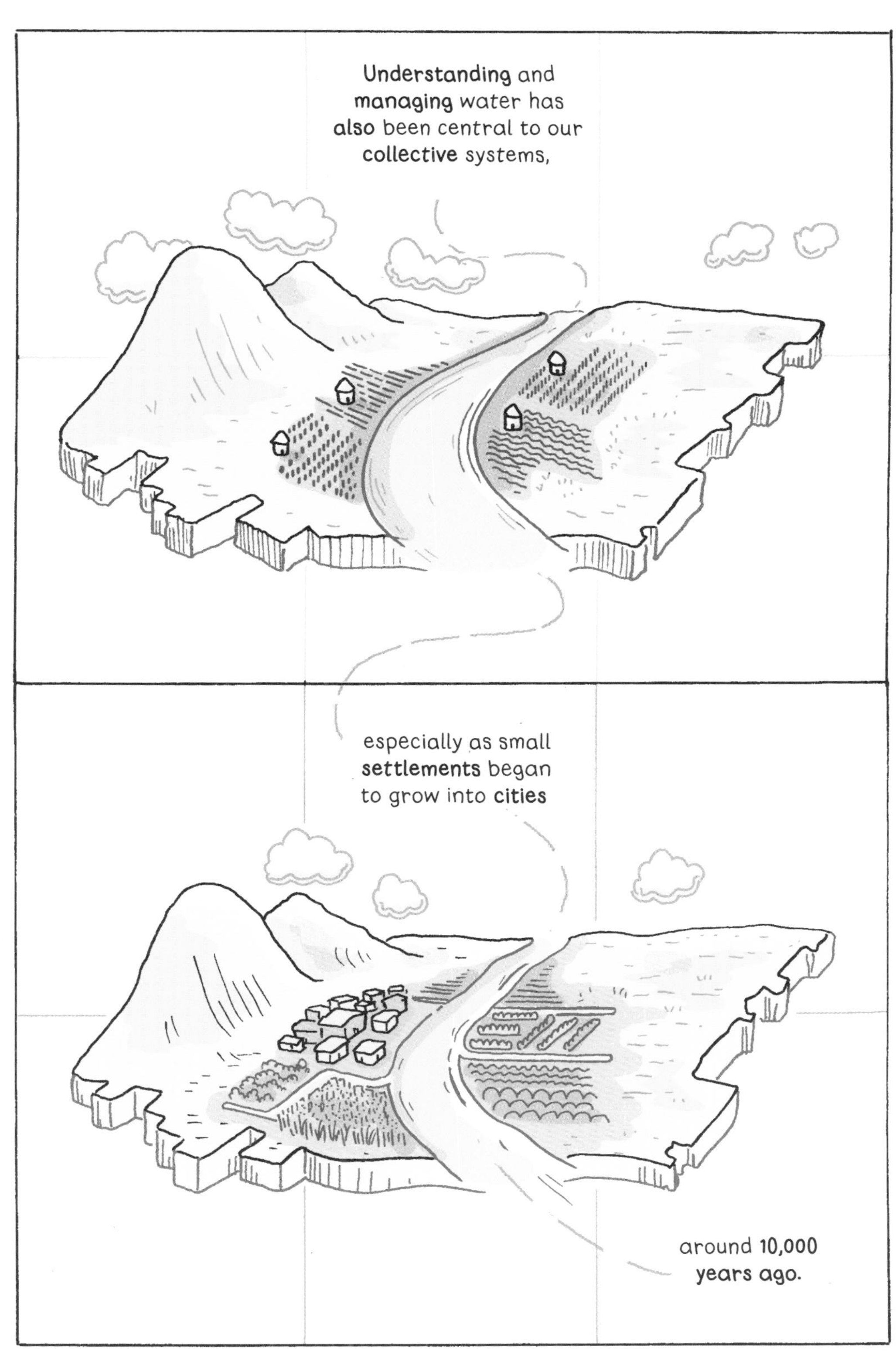
Understanding and **managing** water has **also** been central to our **collective** systems,
especially as small **settlements** began to grow into **cities**
around **10,000 years ago.**

2. Ebbs and Flows, Settlements and Cities

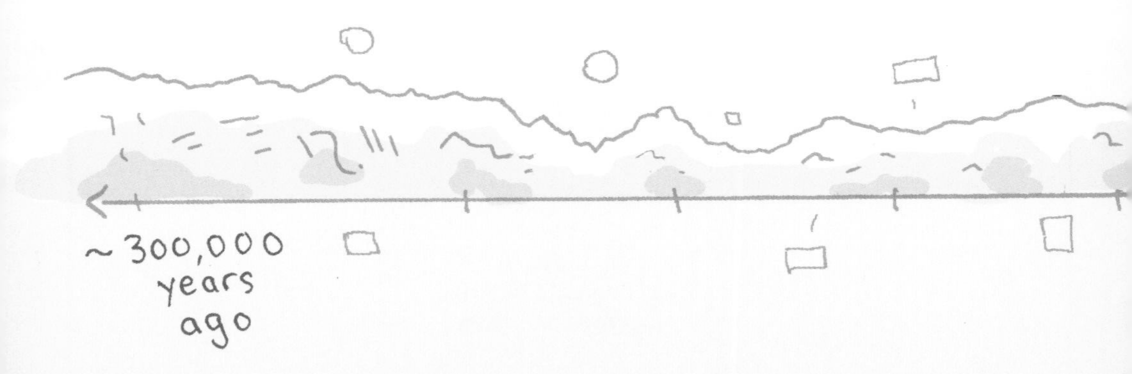

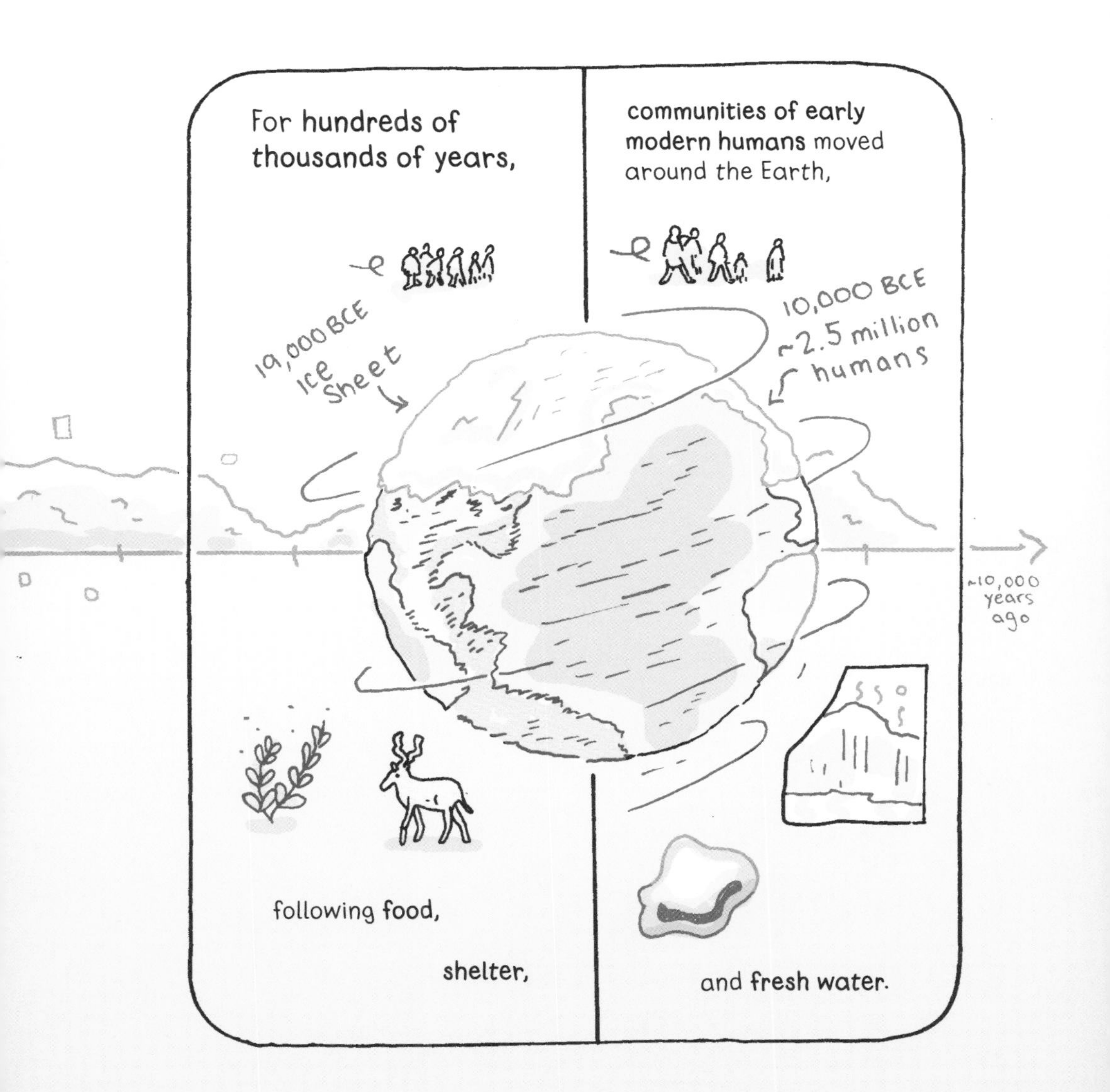
For hundreds of thousands of years,
communities of early modern humans moved around the Earth,
19,000 BCE
Ice Sheet
10,000 BCE
~2.5 million humans
~10,000 years ago
following food,
shelter,
and fresh water.

Over much of the last **100,000 years,**
large ice sheets stretched across the northern part of the planet.

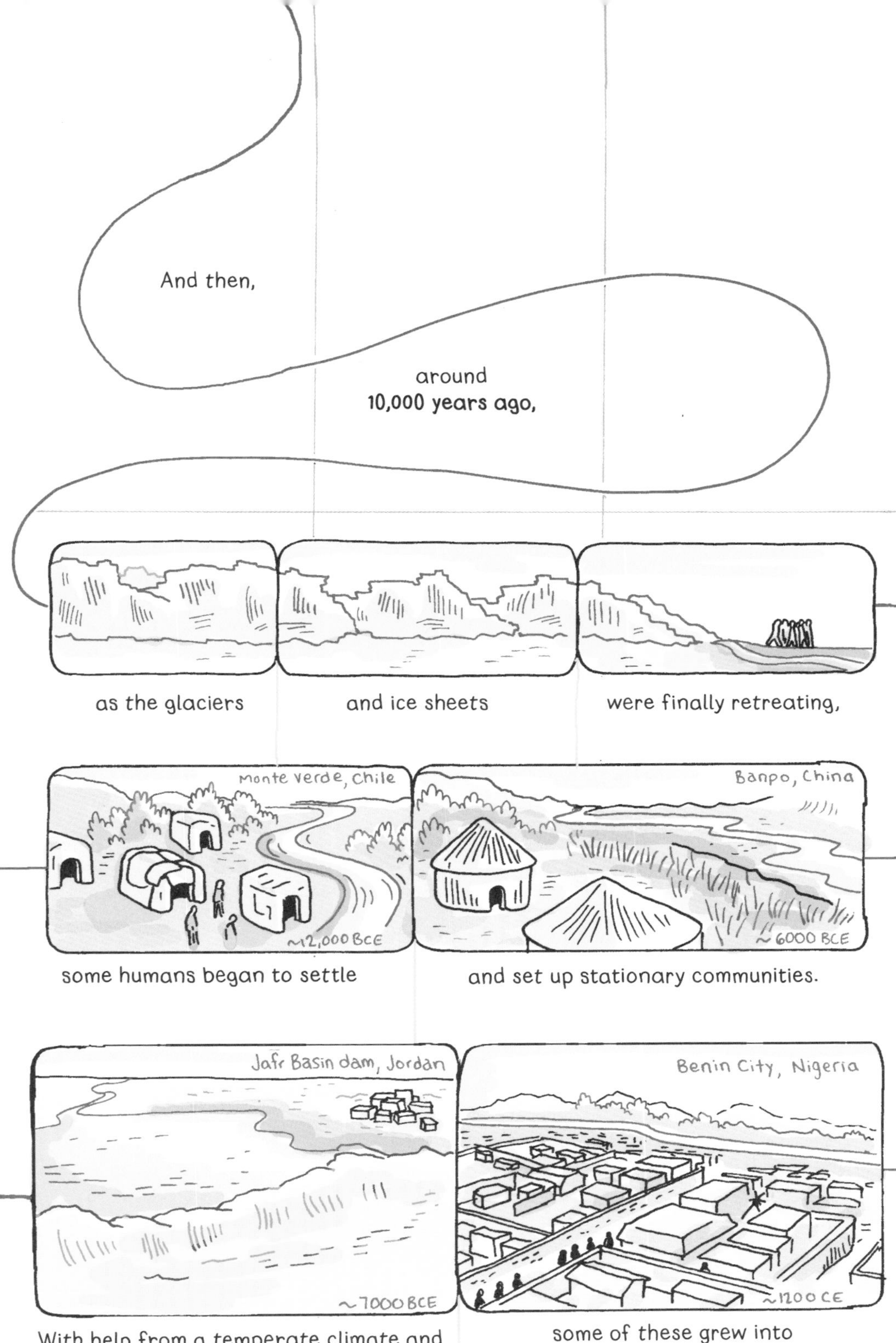
And then,
around
10,000 years ago,
as the glaciers
and ice sheets
were finally retreating,
Monte Verde, Chile
~12,000 BCE
some humans began to settle
Banpo, China
~6000 BCE
and set up stationary communities.
Jafr Basin dam, Jordan
~7000 BCE
With help from a temperate climate and reliable, fresh flowing water,
Benin City, Nigeria
~1200 CE
some of these grew into thriving cities.

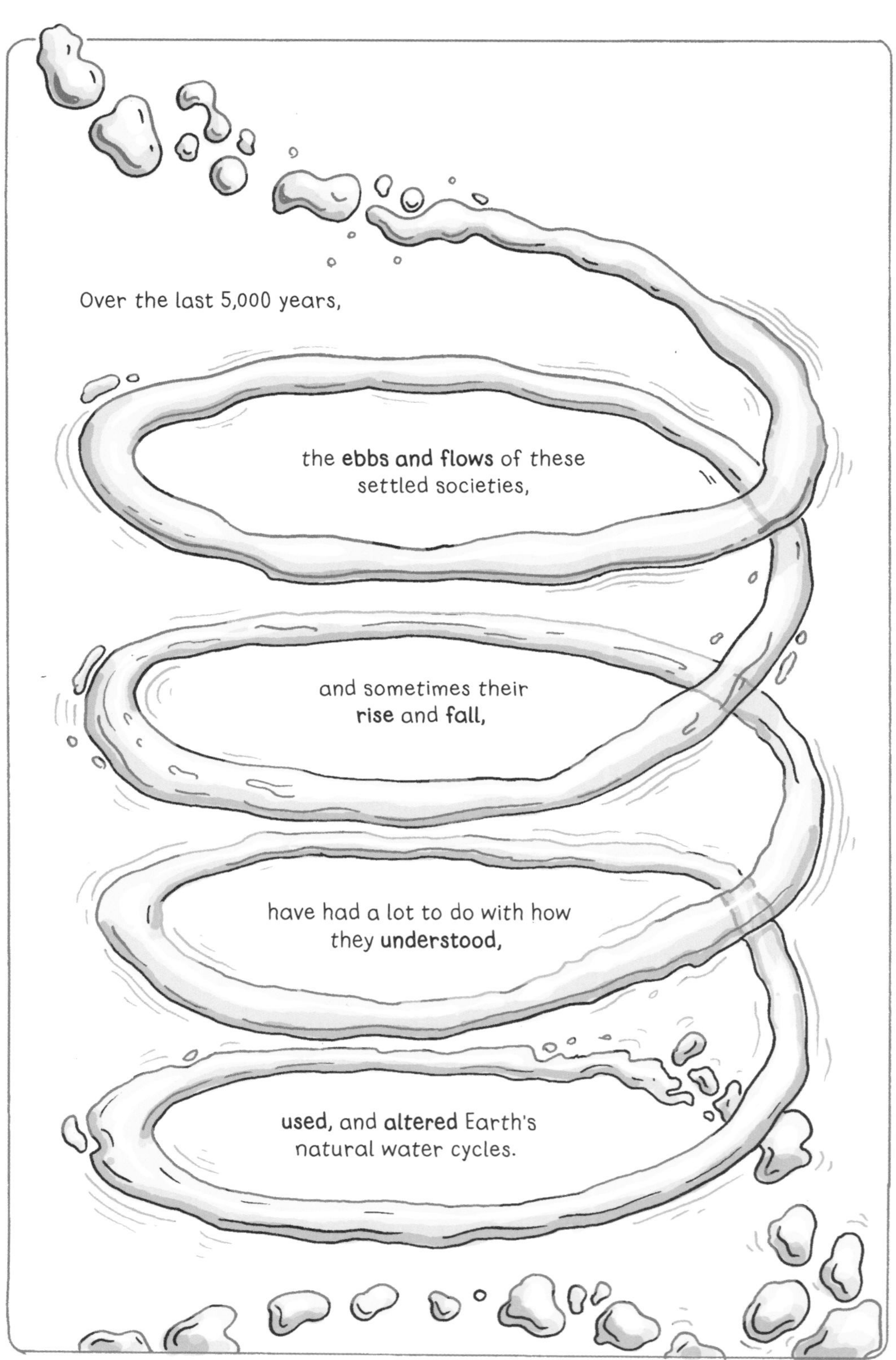
Over the last 5,000 years,
the **ebbs and flows** of these settled societies,
and sometimes their **rise** and **fall,**
have had a lot to do with how they **understood,**
used, and **altered** Earth's natural water cycles.

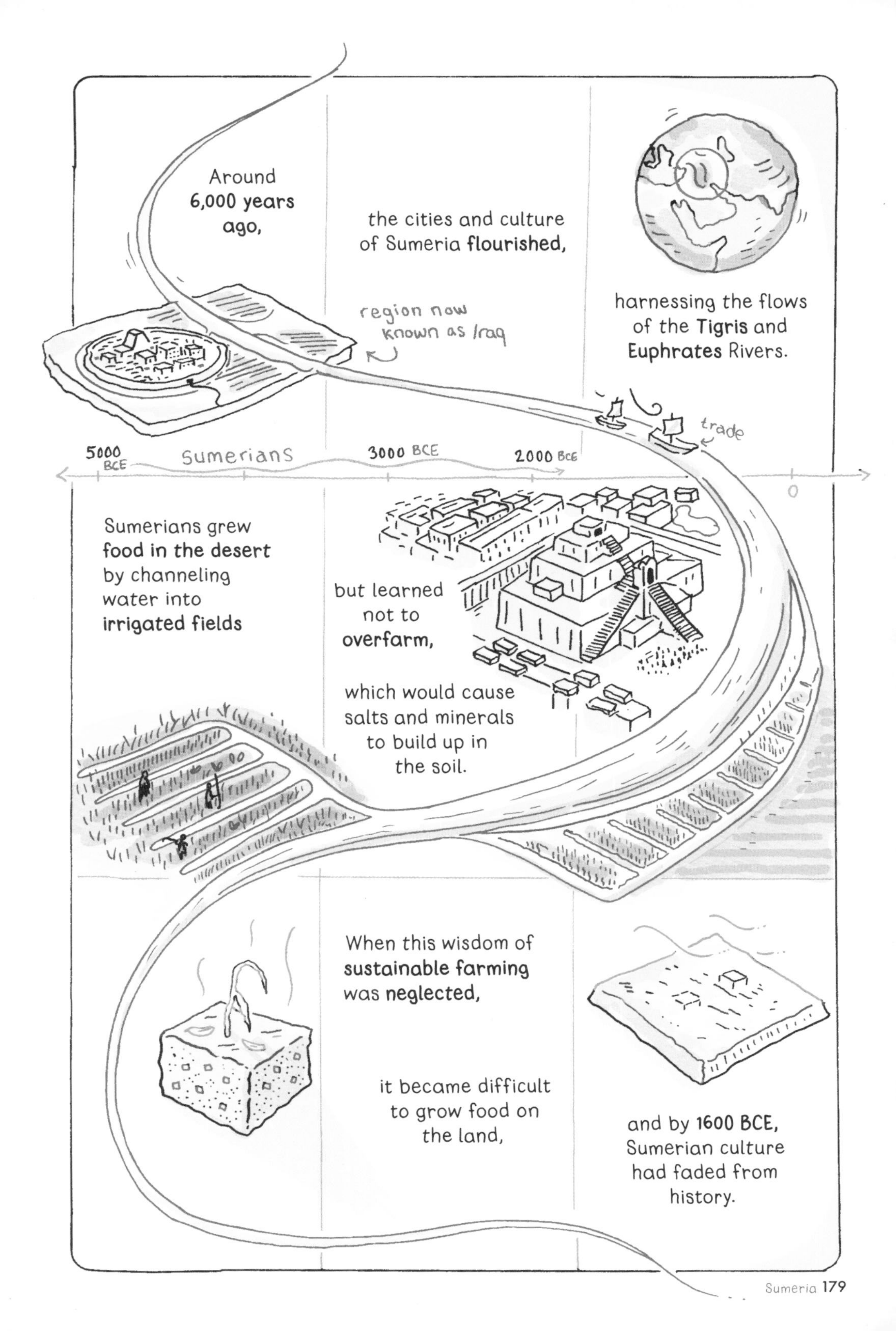
Around
6,000 years ago,
the cities and culture of Sumeria **flourished,**
region now known as Iraq
harnessing the flows of the **Tigris** and **Euphrates** Rivers.
trade
5000 BCE
Sumerians
3000 BCE
2000 BCE
0
Sumerians grew **food in the desert** by channeling water into **irrigated fields**
but learned not to **overfarm,**
which would cause salts and minerals to build up in the soil.
When this wisdom of **sustainable farming** was **neglected,**
it became difficult to grow food on the land,
and by **1600 BCE,** Sumerian culture had faded from history.

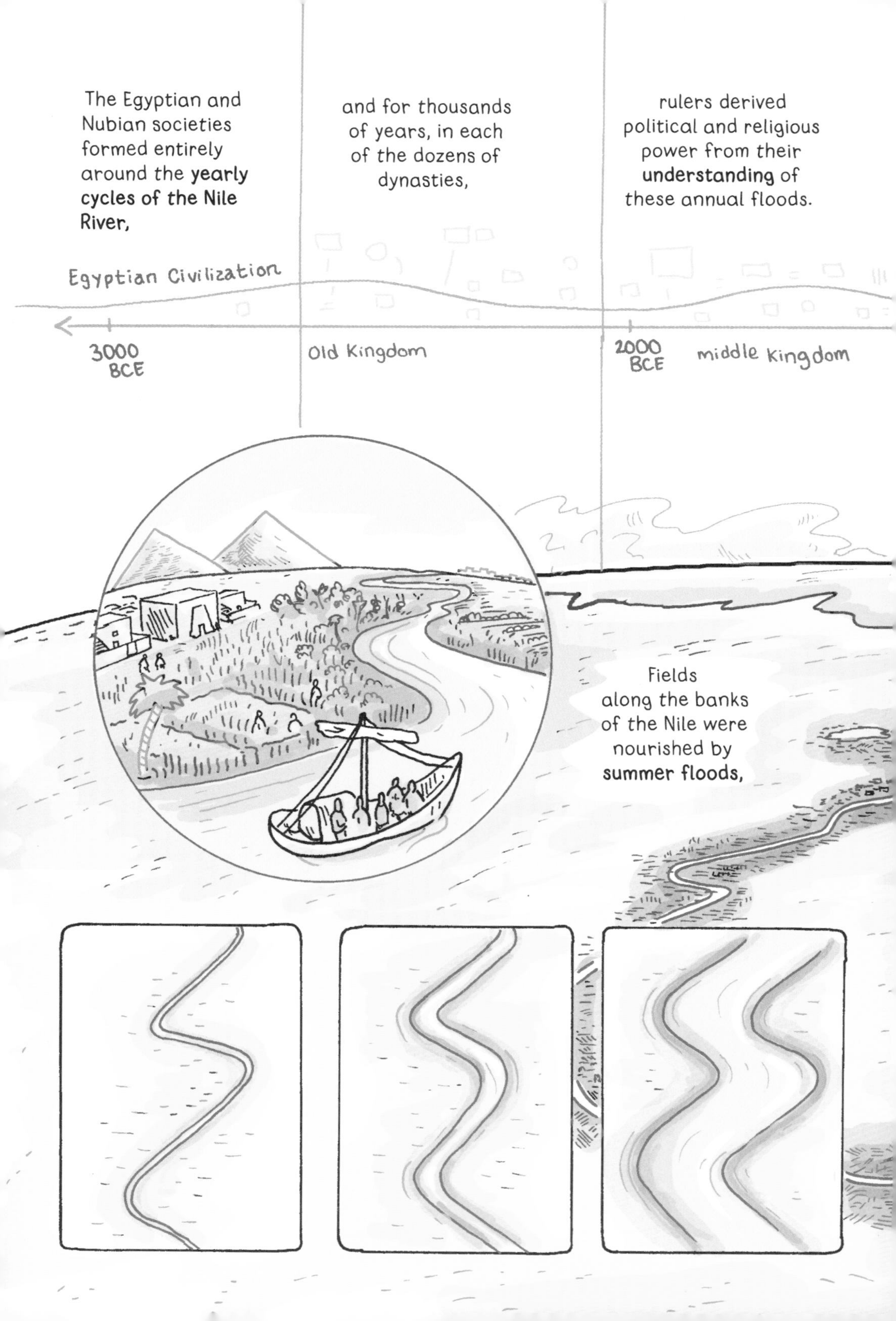
The Egyptian and Nubian societies formed entirely around the **yearly cycles of the Nile River,**
and for thousands of years, in each of the dozens of dynasties,
rulers derived political and religious power from their **understanding** of these annual floods.
Egyptian Civilization
3000 BCE
Old Kingdom
2000 BCE
middle kingdom
Fields along the banks of the Nile were nourished by **summer floods,**

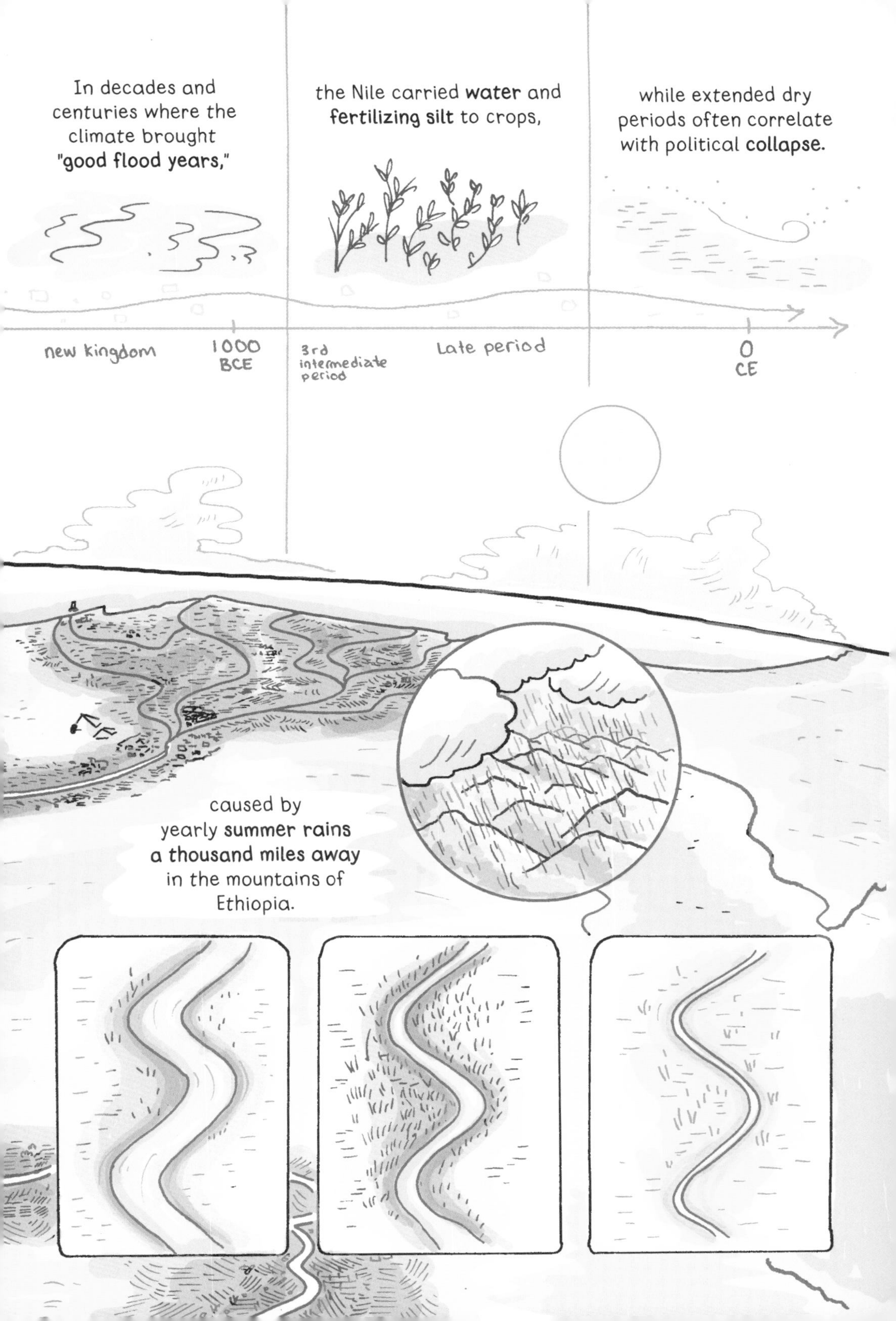
In decades and centuries where the climate brought "good flood years,"
the Nile carried water and fertilizing silt to crops,
while extended dry periods often correlate with political collapse.
new kingdom
1000 BCE
3rd intermediate period
Late period
0 CE
caused by yearly summer rains a thousand miles away in the mountains of Ethiopia.

4,000 miles away, in **East Asia,**

ancient Chinese communities worked to manage the **floods** from **two major river systems,**

China

Yellow

Yangtze

which frequently washed away **dams** and **levees** along the rivers and their tributaries.

700 BCE

272 BCE

Li Bing used a **Taoist** approach to living within these water cycles.

Chinese hydraulic engineer 3rd century BCE

He studied the Min River, a tributary of the Yangtze in the south, for a way to work **within** its flows,

and workers built a series of diversions and weirs to **split** the river at a key bend.

rocks piled in bamboo cages

The waterworks **redirected** excess water to prevent floods

and **irrigated** hundreds of square miles of farmland,

a marvel of engineering that still works today **over 2,000 years later.**

Dujiangyan Irrigation System

Min River

In northern China, it was a **regular lack of water** that posed a dire threat.

Around **600 CE,** using a more forceful approach,

China

Yellow

Yangtze

the Sui dynasty ordered the creation of the **longest artificial river** in the world.

700 CE

600 CE

Using the labor of over **3 million people,** the dynasty cut a **canal** through the land,

connecting **existing canals** with the Yellow and Yangtze Rivers,

creating a **network** to move **water, food,** and **trade** from south to north.

Complex and frequently damaged from **floods, warfare,** and **lack of maintenance,**

Mud and silt have to be continually cleared out.

the many sections of the **Grand Canal** have been **expanding** and **contracting** for **1,500 years,**

alongside the ebbs and flows of each dynasty.

In 312 BCE, Romans built a small artificial river of a different type.
312 BCE
Unable to pull enough water from the polluted Tiber River,
they built an underground aqueduct to supply the capital with fresh water from the countryside.
moves by gravity over 10 miles
Aqueducts had been built before,
Iranian Qanats
Assyrian Aqueducts
Greek Pipes
but the Romans are remembered for building them to new heights
40-60 CE Pont du Gard
and providing public water for essentials like drinking and bathing.
used lead pipes
Rome also used water for decoration and social life,
water warmed from nearby furnaces
and as a source of entertainment
like filling stadiums
to stage naval battles
for a rapidly growing population at the heart of an empire.

Around 500 years later, in **226 CE,** Rome added its **eleventh** aqueduct,
and for the first time,
an estimated population of **around 1 million people** lived in a single city.
food imported from the Nile
700 CE
But Rome's waterworks didn't last—
in **537 CE,** as its empire was falling,
invading armies cut all but one of Rome's water lines, and leaders abandoned the city.
Rome's population shrank to a **fraction** of its former size—
it would take **1,000 years** before many of the aqueducts were restored
and until the **1930s** for the city to regain its former population.

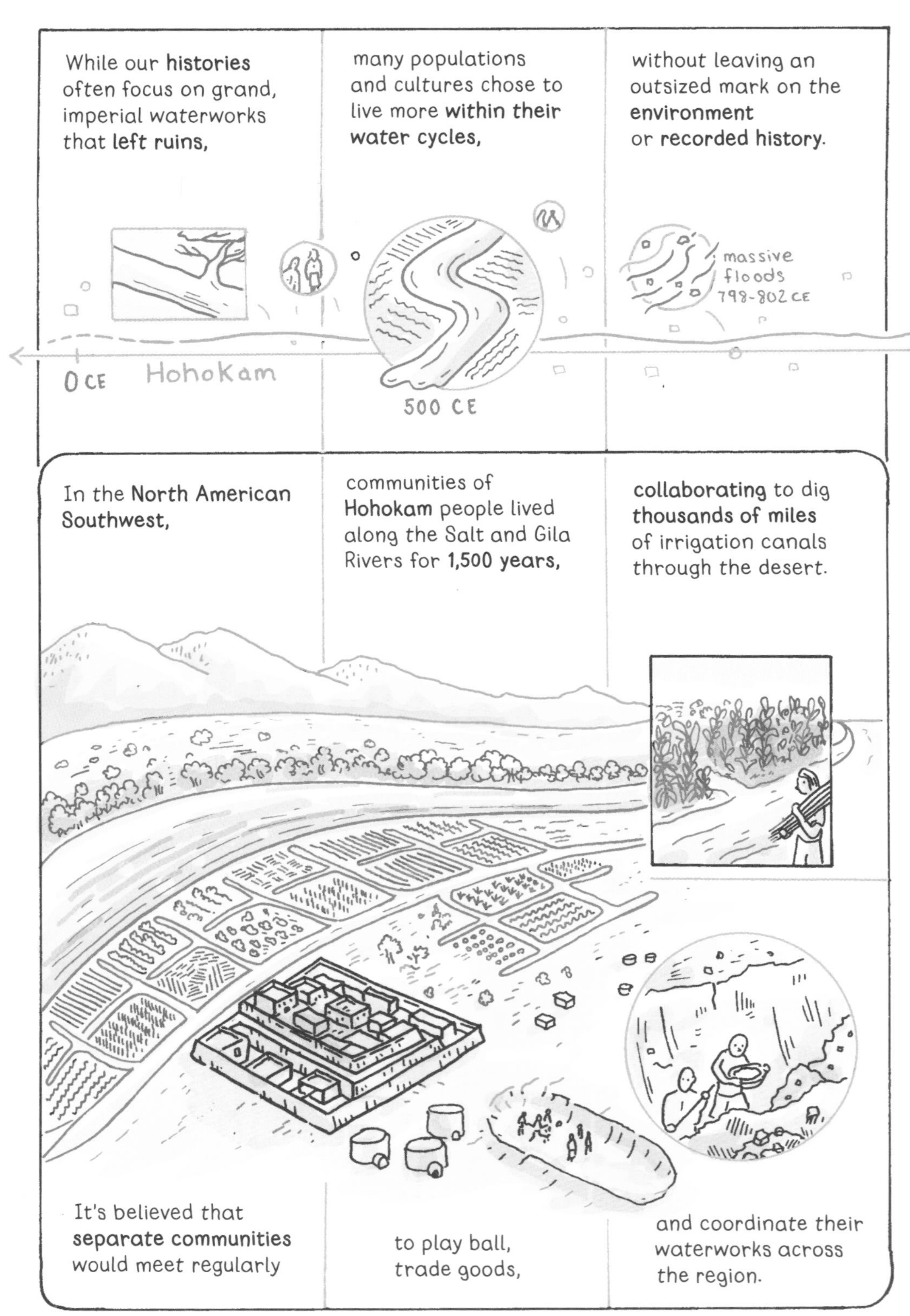
While our **histories** often focus on grand, imperial waterworks that **left ruins,**
many populations and cultures chose to live more **within their water cycles,**
without leaving an outsized mark on the **environment** or **recorded history**.
0 CE
Hohokam
500 CE
massive floods 798-802 CE
In the **North American Southwest,**
communities of **Hohokam** people lived along the Salt and Gila Rivers for **1,500 years,**
collaborating to dig **thousands of miles** of irrigation canals through the desert.
It's believed that **separate communities** would meet regularly
to play ball, trade goods,
and coordinate their waterworks across the region.

Our histories are based on what's **left behind,**
by what is **prioritized** and **collected** by those recording and retelling it.
Cultures that lived in **harmony** with nature are often less likely to be studied today.
drought 1322-59
1500 CE
1380-82 massive floods
1867 Phoenix, AZ
2000 CE
Sometime in the **1300s,**
likely due to a difficult century of **floods and droughts,**
Hohokam apparently abandoned their settlements,
and most evidence of the culture vanished.
Centuries later, in the 1860s,
FUTURE HOME SWILLING CANAL CO
European colonizers noticed the ditches and **redug** canals along the Hohokam routes,
which are used today for water by **Phoenix, Arizona.**

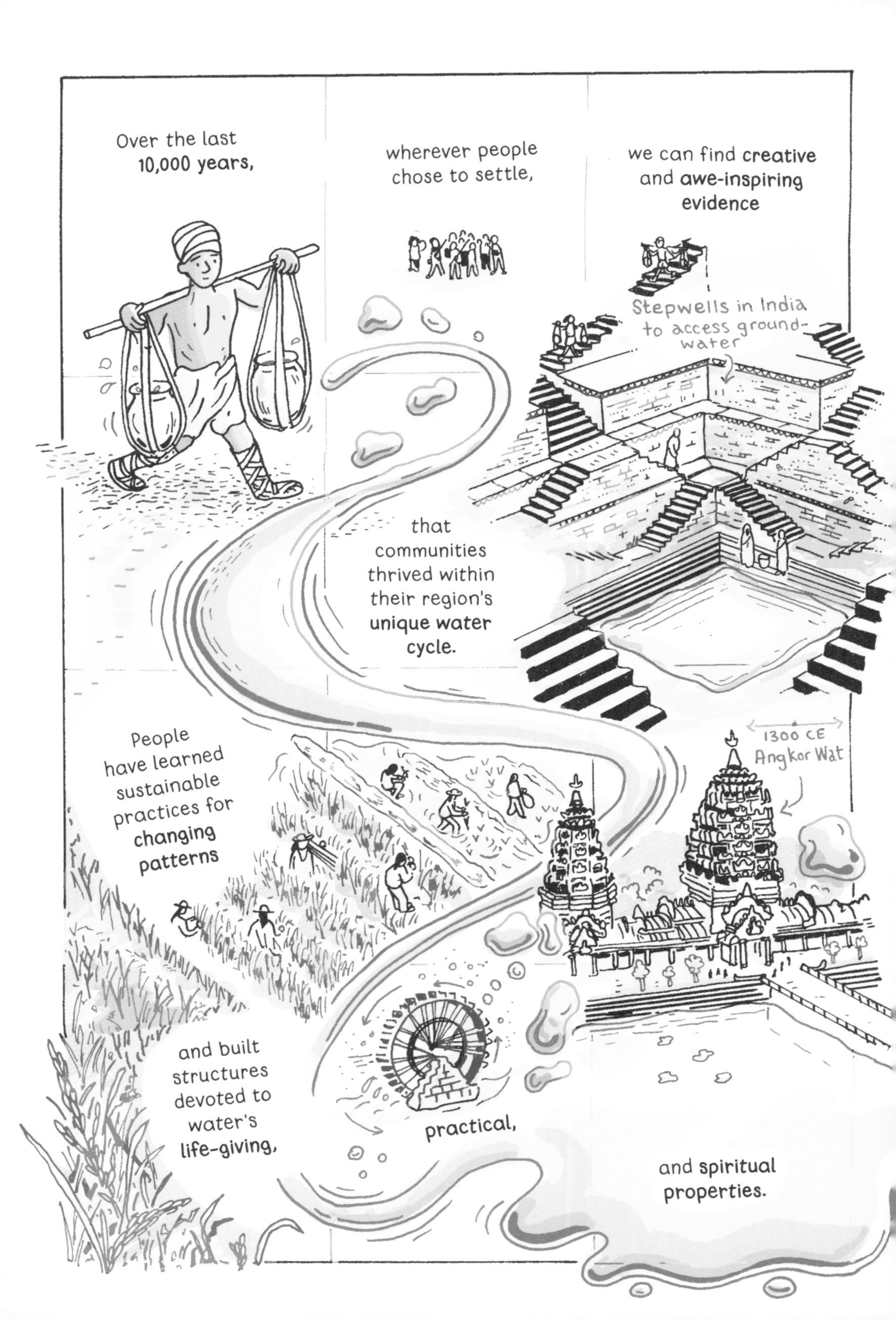
Over the last **10,000 years,**
wherever people chose to settle,
we can find **creative** and **awe-inspiring evidence**
Stepwells in India to access ground-water
that communities thrived within their region's **unique water cycle.**
1300 CE
Angkor Wat
People have learned sustainable practices for **changing patterns**
and built structures devoted to water's **life-giving,**
practical,
and **spiritual properties.**

Petra (Nabataeans)
200 BCE
700 CE
rare desert rainwater stored in cisterns
These monuments and systems often **outlast** the communities that made them
and remind us that regardless of what we've built—
Tikal (Maya)
600 BCE
1000 CE
adapting to **changing water cycles** can be a matter of life and death.
reservoirs faced drought and pollution
Wisdom about **our place in the water cycle** is just as important today
but considered much less,
after **populations grew**
and cities **industrialized**
over the last **200 years.**

3. Water in the City

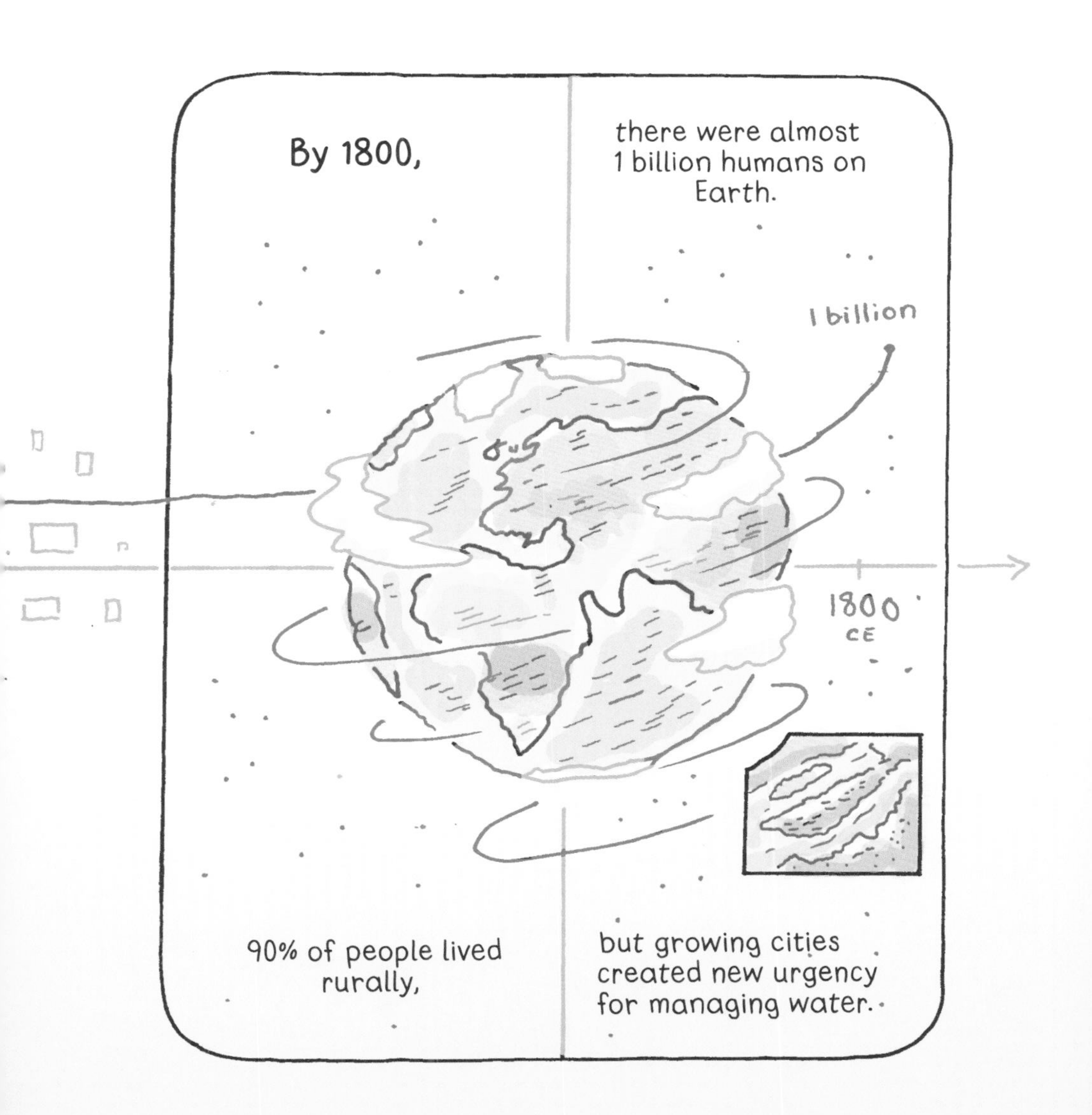
By 1800,
there were almost 1 billion humans on Earth.
1 billion
1800 CE
90% of people lived rurally,
but growing cities created new urgency for managing water.

In 1600, precolonial **Mannahatta Island** was a collection of **55 ecosytems,**
supporting **wildlife** and **communities of humans**
who had lived there for **thousands of years.**
Lenape
The edge of a massive glacier 18,000 years ago
over 2,000 feet high
Dutch settlers colonized the tip of the island, and it was renamed **New York** in 1664 by the British.
Wall St.
1660
But the water of the Hudson and East River was **too salty to drink,**
1776
and **fresh water** became **scarce** as the city grew and settlers polluted the island's ponds.
1800
People set up commercial "tea pumps" in places where groundwater tasted clean,
and water was sold at high prices and carried into the homes of those who could afford it,
an **unsustainable** system made more urgent by frequent fires that raged in the city.

The city tried to build a drinking water system by hiring the **Manhattan Company,** created by **Aaron Burr.**

But the company hardly did any work besides burying some leaky pipes

because it was mostly a front for Burr's new bank.

Like most growing cities throughout history,

New York found its solution by **diverting distant sources** of **clean water through aqueducts,**

moving it **forty** miles from the Croton River to a **reservoir** on the island.

The arrival of **fresh water** in New York in 1842 was a cause of jubilant celebration

and marked the start of an **expansive drinking water system**

that now channels a **billion gallons** daily from **nineteen reservoirs** and **three lakes.**

In addition to the challenge of getting water into a city,
urban residents also quickly realized the importance of getting it out.
Watch your step!
SPLASH
In the 18th and 19th centuries, growing European and American cities were dirty and disease-ridden.
While natural earth like forests and fields can absorb water,
developed land creates pools of stagnant water
and runoff that often flows into the city's rivers.
In 1712, the satirist Jonathan Swift described:
"sweepings from butcher's stalls,
dung,
... drowned puppies,
stinking sprats all drenched in mud,
dead cats and turnip tops"
. . . seen flowing down the street as sewers and cesspools failed to keep up with population growth.

London's river, the Thames, had become the city's **sewer.**
As the city grew, and residents created more **human** and **industrial waste,**
they drained it out of the city using the **flows of the river**
without understanding the **danger** of mixing **water and sewage.**
When **cholera pandemics** began spreading rapidly in the 1800s,
they were difficult to stop, in part because people still believed sickness was caused by **smells**—
The bad air!
and they didn't know about the **bacteria** living in the water they drank.
During a particularly warm summer in **1858,** the unbearable smell of the Thames led to action.
When even the city's **wealthy** and **Parliament** were subjected to the stench,
they began drafting plans for an **overhaul.**
1858 – "The Great Stink"

18th- and 19th-century Paris was not in much better shape.
Tout à la rue!
splash
While human feces were stored in **underground pits or tanks,**
people often washed excess water and waste down narrow streets to the river.
splash
As Paris's restive population swelled past **1 million** by **1850,**
Louis-Napoléon ordered a forced reconstruction of the city's urban space,
BOOM
demolishing **entire neighborhoods** and widening city streets.
In the redesign, **fresh water** was channeled from the countryside for drinking,
and river water was pumped up to wash waste from the streets,
shhhh
all of which was moved by a massive network of underground sewers.
(Later carried power and communication too)

Parisians built a magnificent and extremely over-budget sewer system that carried nearly all the city's waterworks,
meant to be toured and admired
Haussmann, architect
but not **solid human waste,**
They want to put **what** in my sewers?
which was still moved to farms for **use as fertilizer** by dozens of teams of **night soil men.**
Londoners created a more practical **combined** system,
protecting the water supply for its rapidly growing population
by dumping the waste farther downriver.
1860s
New designs for urban waterworks spread around the world, creating safer, cleaner cities,
built brick by brick
but **rivers** remained essential for their ability to move industrial pollution somewhere else,
BOSTON
TOKYO
especially as businesses found new ways to use **water for work.**

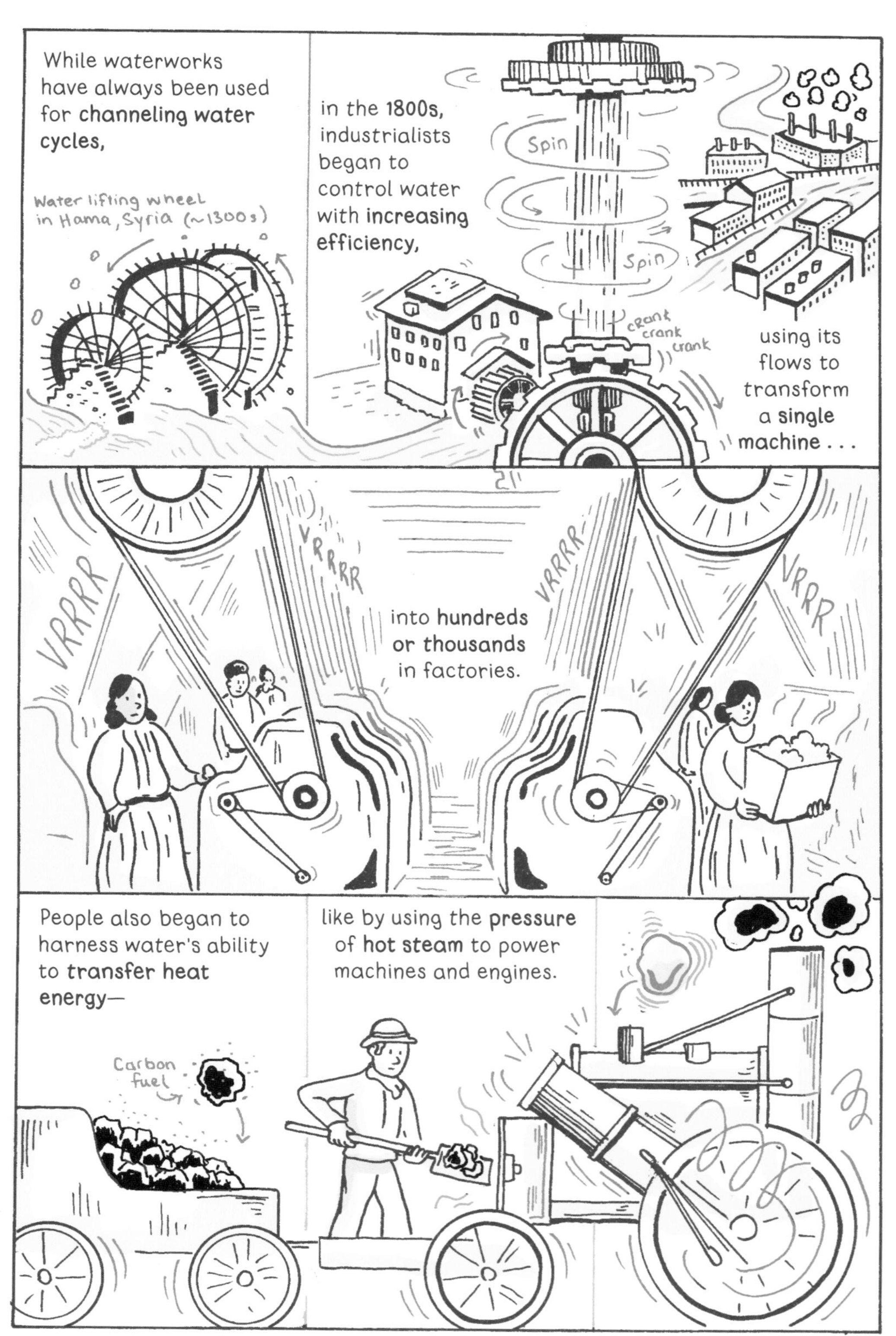
While waterworks have always been used for **channeling water cycles,**
Water lifting wheel in Hama, Syria (~1300s)
in the **1800s,** industrialists began to control water with **increasing efficiency,**
Spin
Spin
crank crank crank
using its flows to transform a **single machine . . .**
VRRRR
VRRRR
VRRRR
VRRR
into **hundreds or thousands** in factories.
People also began to harness water's ability to **transfer heat energy**—
Carbon fuel
like by using the **pressure** of **hot steam** to power machines and engines.

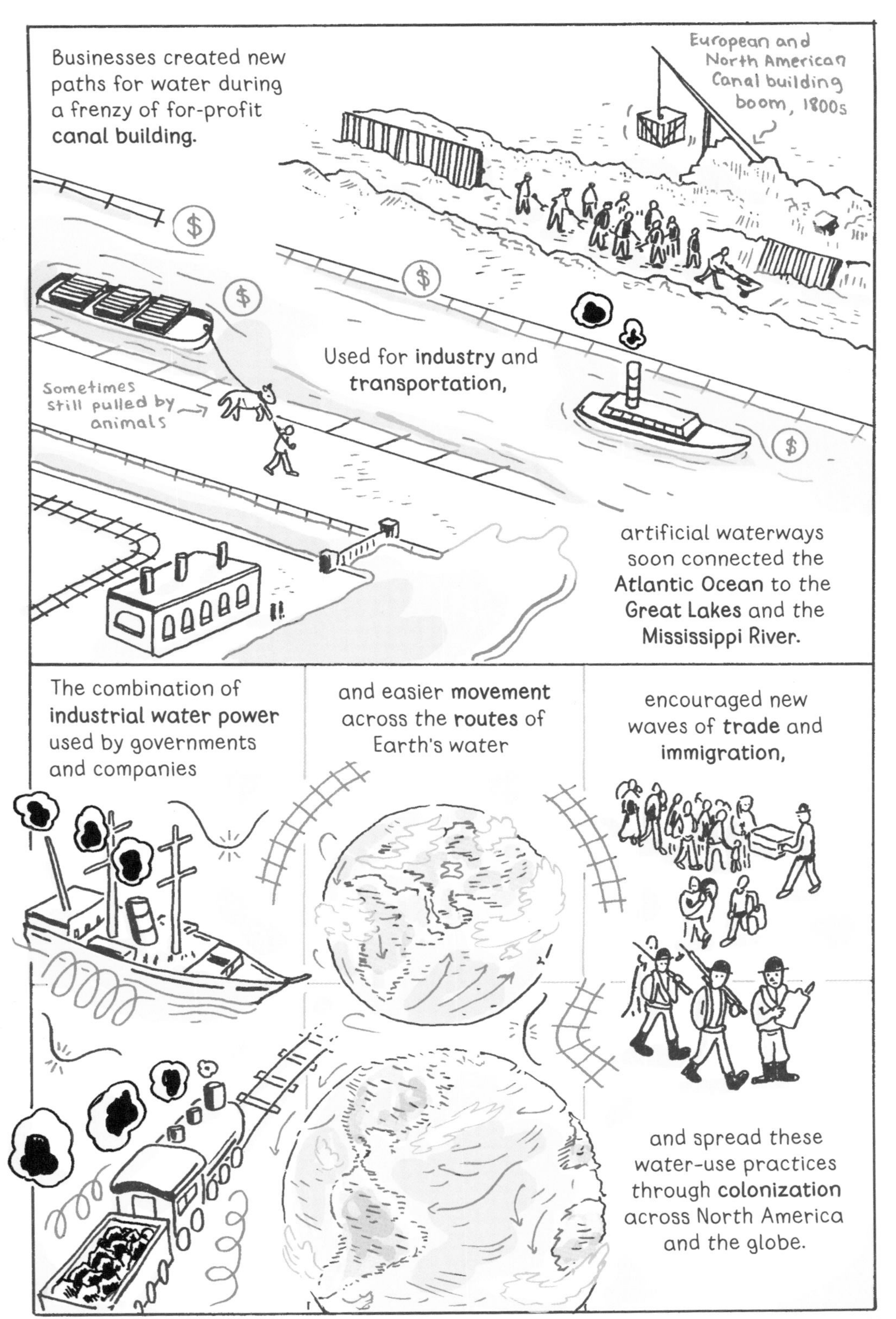
Businesses created new paths for water during a frenzy of for-profit **canal building.**
European and North American Canal building boom, 1800s
$
$
$
$
Used for **industry** and **transportation,**
Sometimes still pulled by animals
artificial waterways soon connected the **Atlantic Ocean** to the **Great Lakes** and the **Mississippi River.**
The combination of **industrial water power** used by governments and companies
and easier **movement** across the **routes** of Earth's water
encouraged new waves of **trade** and **immigration,**
and spread these water-use practices through **colonization** across North America and the globe.

4. Claiming Land, Claiming Water

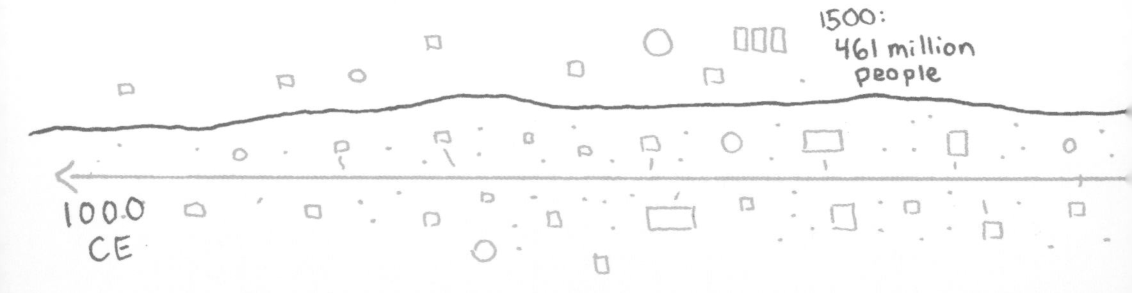

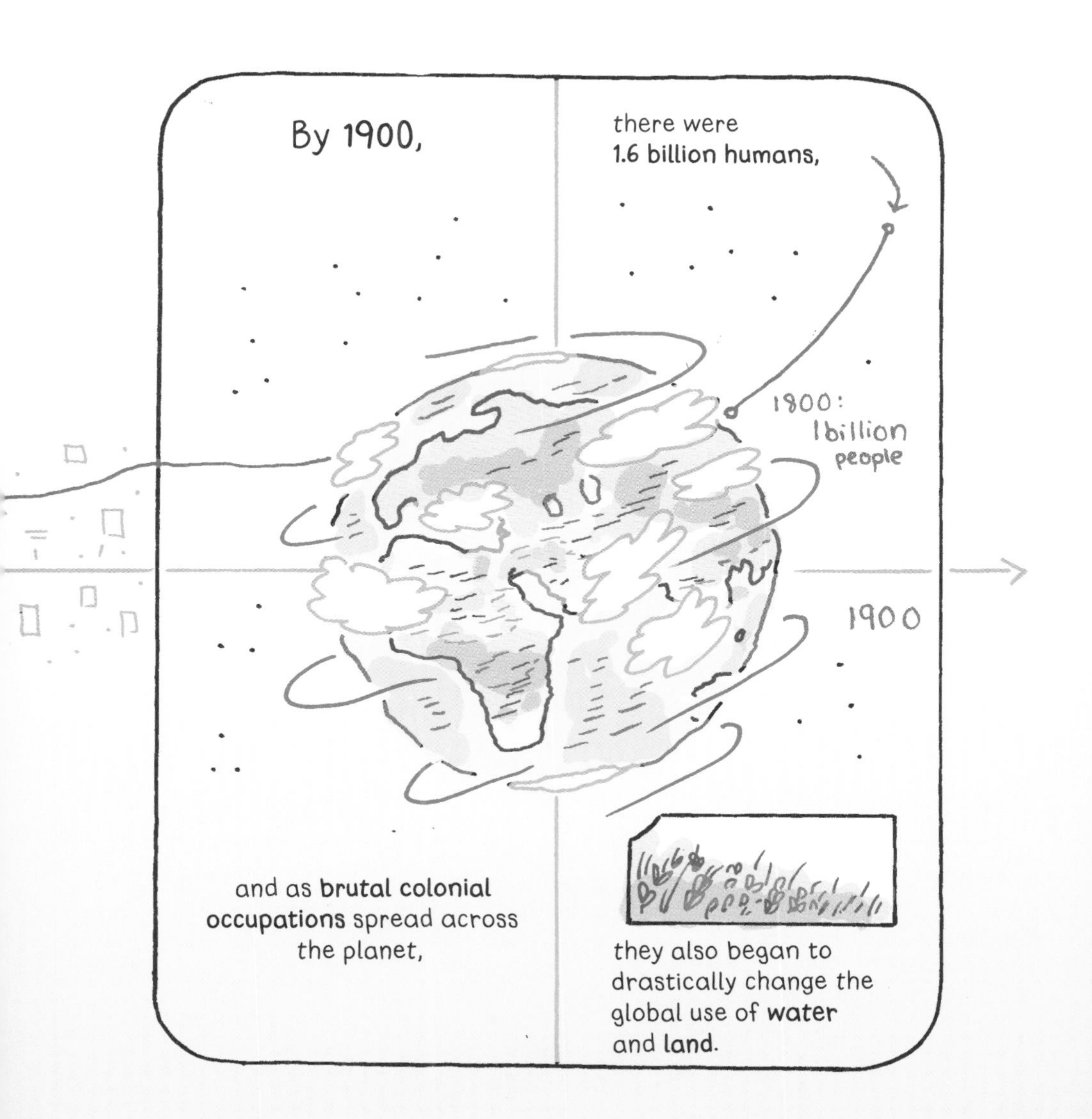
By 1900,
there were
1.6 billion humans,
1800:
1 billion
people
1900
and as **brutal colonial occupations** spread across the planet,
they also began to drastically change the global use of **water** and **land**.

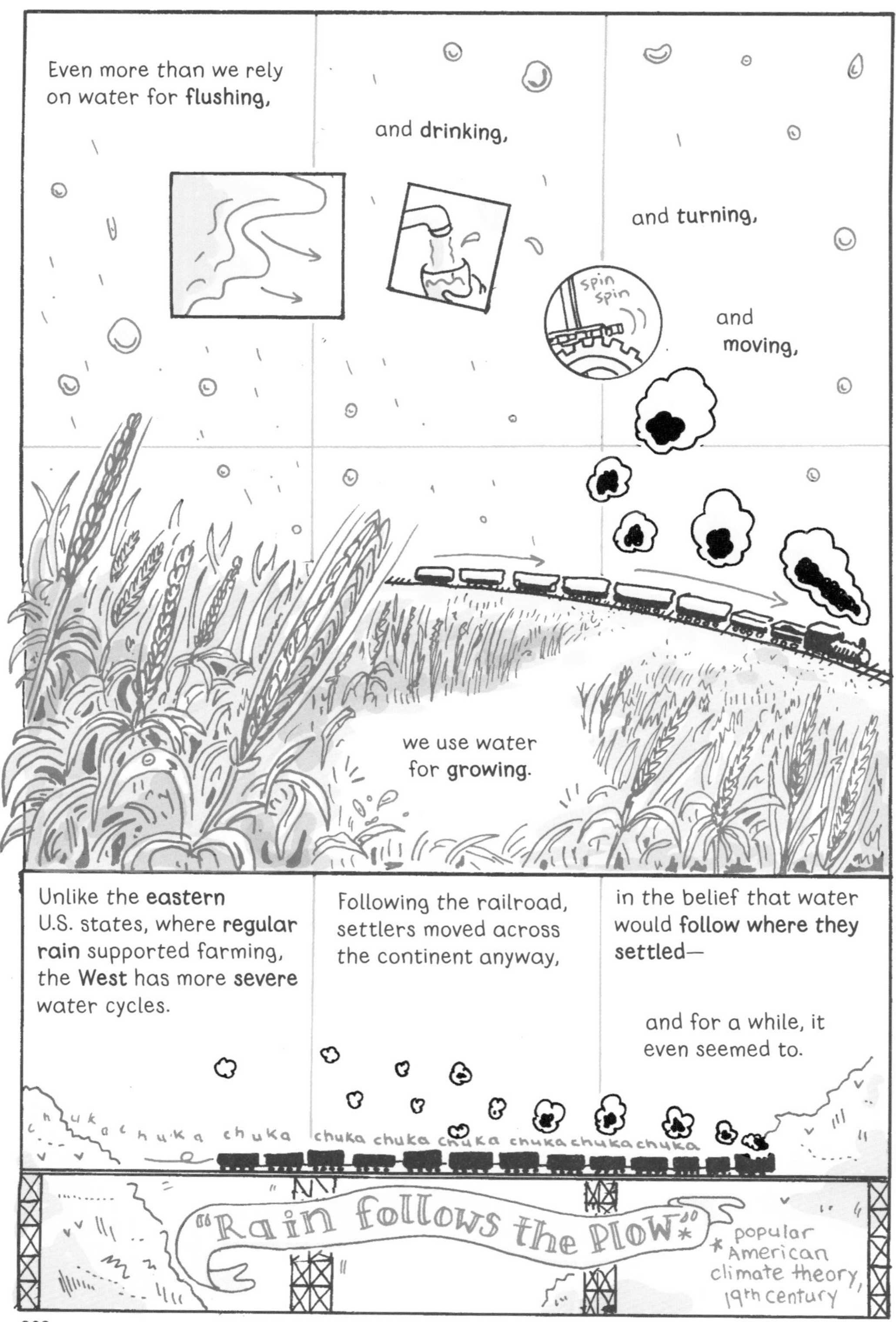

Even more than we rely on water for **flushing,**
and **drinking,**
and **turning,**
spin spin
and **moving,**
we use water for **growing.**
Unlike the **eastern** U.S. states, where **regular rain** supported farming, the **West** has more **severe** water cycles.
Following the railroad, settlers moved across the continent anyway,
in the belief that water would **follow where they settled**—
and for a while, it even seemed to.
chuka chuka chuka chuka chuka chuka chuka chuka chuka chuka
"Rain follows the Plow"*
*popular American climate theory, 19th century

Settlers overran and forced out the **Native American** people of the **Great Plains**, slaughtering their food source, the buffalo,
and **plowing** over the **vast region** in an ever-increasing effort to maximize harvests of cash crops like wheat,
a practice that continued until the rains stopped.
The resulting **Dust Bowl** of the 1930s was an **environmental disaster** of our own making,
where the newly plowed, drought-dry dirt was whipped up by winds into **dust storms**,
creating a lifeless, **desolate** landscape for much of the following decade.
Some settlers stayed, receiving help from the New Deal and waiting for rains to return,
while others continued farther west,
where they hoped water would be more reliable.

Many of these climate refugees arrived in **California in the 1930s,**
where the U.S. was building one of the world's largest water projects—
a **400-mile network** to control water in **California's Central Valley.**
SIERRA NEVADA MOUNTAINS
Sacramento
Oakland
San Francisco
Los Angeles
Pacific Ocean
The **Central Valley Project** uses pumps and canals
to move water from the **snowmelt** of the **Sierra Nevada mountain range**
to farmers and their crops in the harsh valley.
Subsidized water was initially provided to help small farmers struggling with droughts,
though much of the water has found its way to corporate farms—
pick
pick
pick
which turn this redirected water into cash crops.

But in order to supply massive schemes like this with enough water,

the U.S. government and states began forceful transformations of the continent's waterways

by building walls across the rivers.

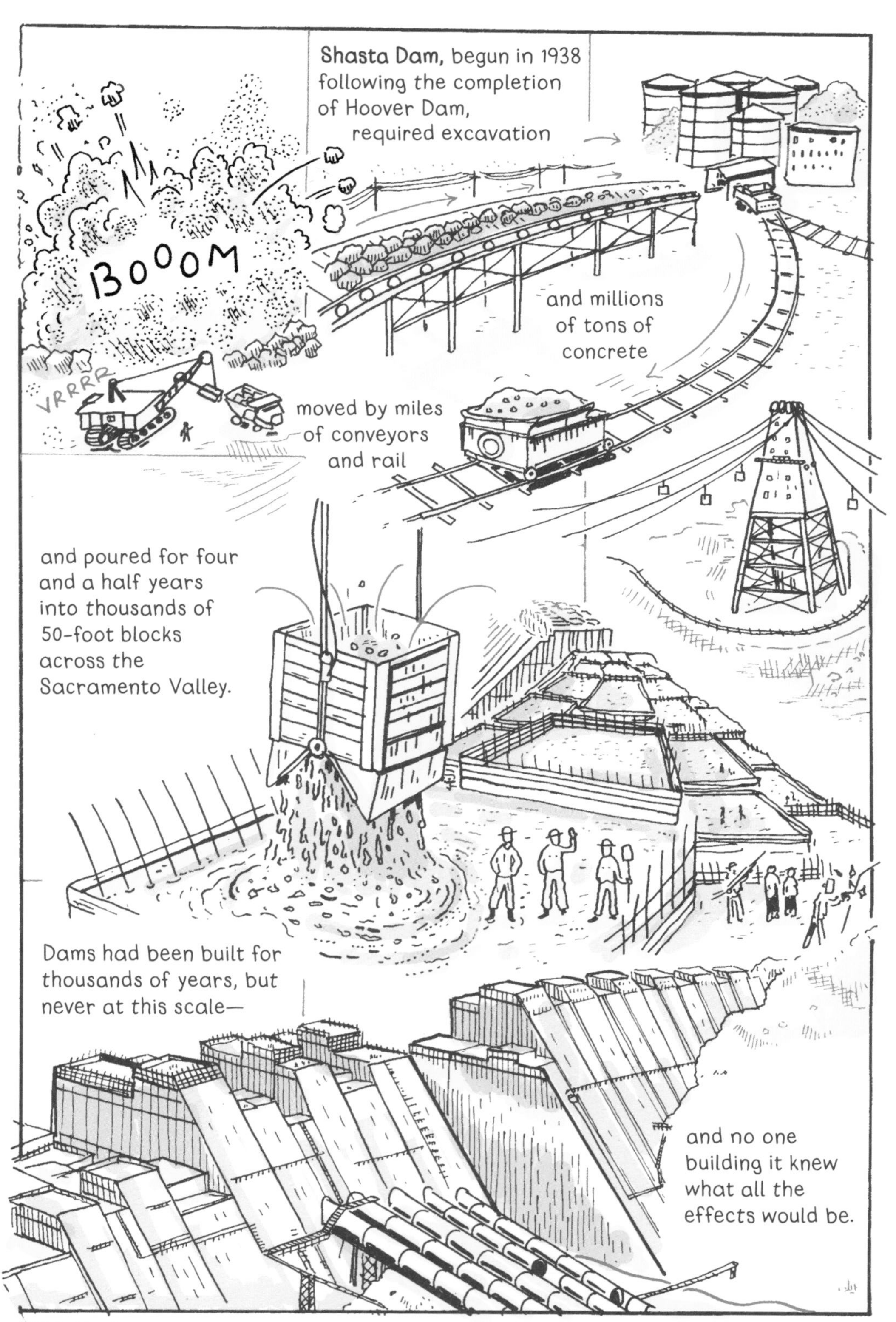
Shasta Dam, begun in 1938 following the completion of Hoover Dam, required excavation
BOOOM
VRRRR
and millions of tons of concrete
moved by miles of conveyors and rail
and poured for four and a half years into thousands of 50-foot blocks across the Sacramento Valley.
Dams had been built for thousands of years, but never at this scale—
and no one building it knew what all the effects would be.

The decision to dam a river allowed the government to disrupt water cycles for a **number of purposes**—
controlling a river's **natural floods,**
using the falling water to **generate electricity,**
CLK
and channeling it into systems for **drinking** and **growing,**
VRRRR
while also creating a vast **reservoir** of water by flooding **tens of thousands of acres of land.**

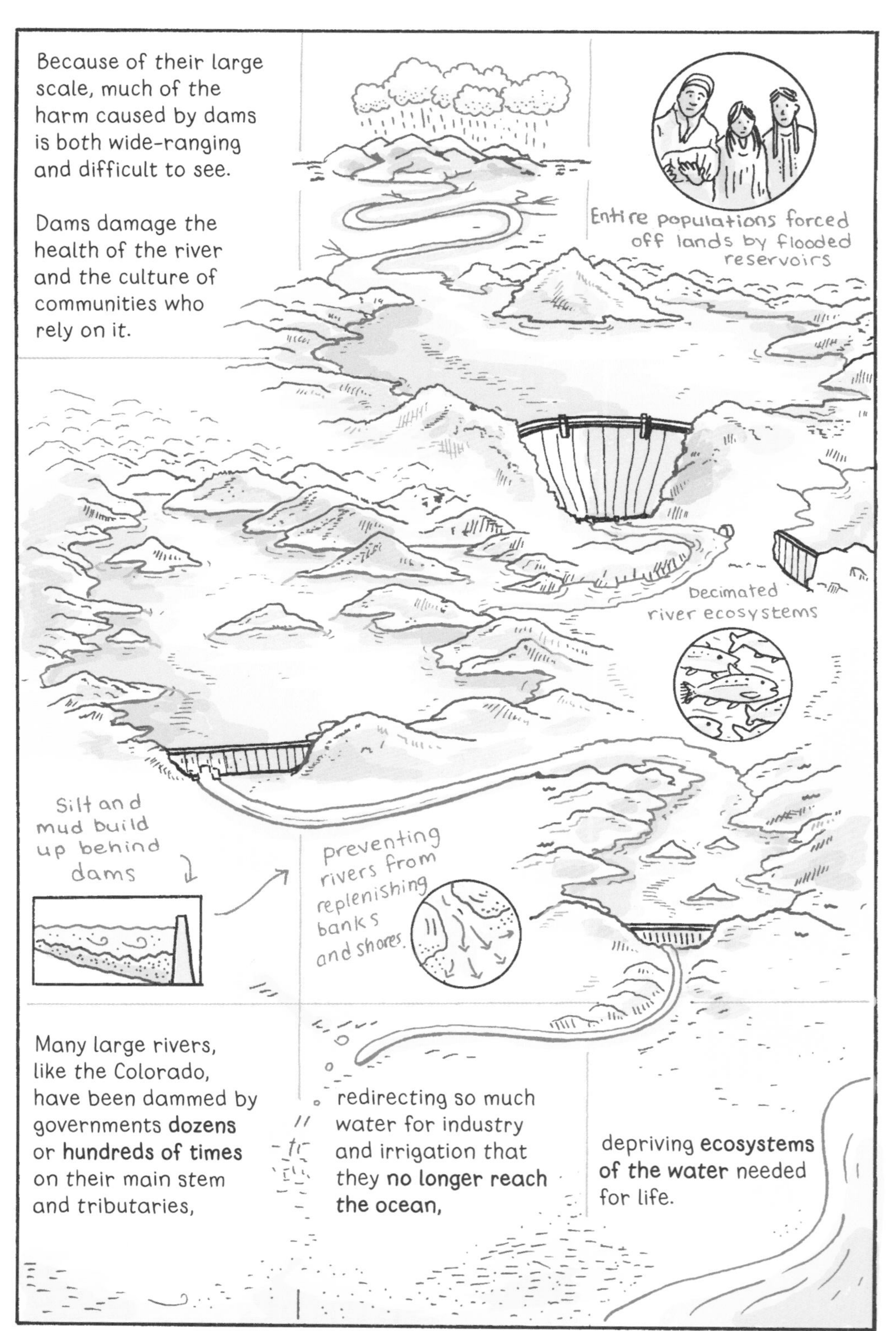
Because of their large scale, much of the harm caused by dams is both wide-ranging and difficult to see.
Dams damage the health of the river and the culture of communities who rely on it.
Entire populations forced off lands by flooded reservoirs
Decimated river ecosystems
Silt and mud build up behind dams
preventing rivers from replenishing banks and shores.
Many large rivers, like the Colorado, have been dammed by governments **dozens** or **hundreds of times** on their main stem and tributaries,
redirecting so much water for industry and irrigation that they **no longer reach the ocean,**
depriving **ecosystems of the water** needed for life.

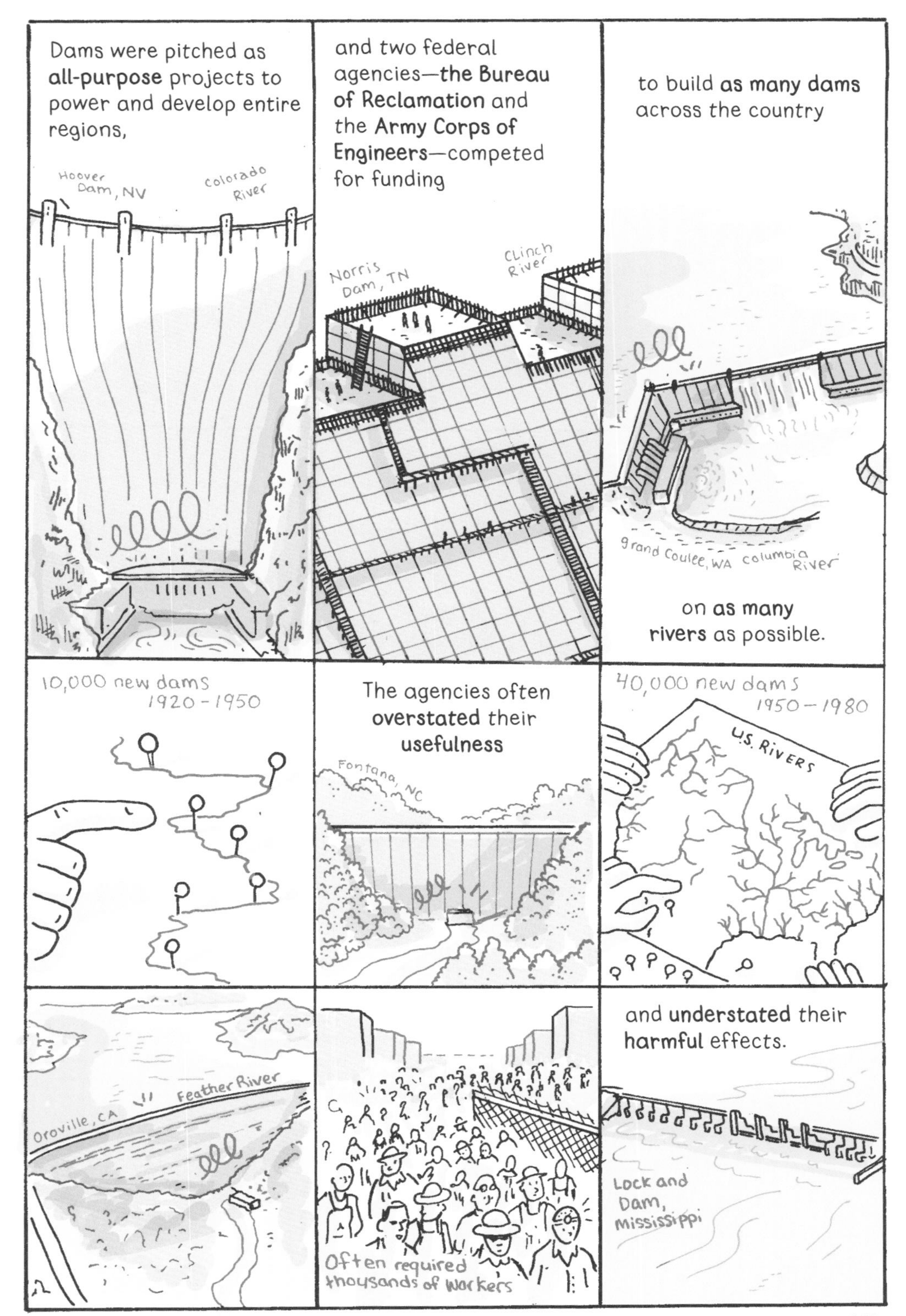
Dams were pitched as **all-purpose** projects to power and develop entire regions,
Hoover Dam, NV
Colorado River
and two federal agencies—**the Bureau of Reclamation** and the **Army Corps of Engineers**—competed for funding
Norris Dam, TN
Clinch River
to build **as many dams** across the country
Grand Coulee, WA
Columbia River
on **as many rivers** as possible.
10,000 new dams 1920-1950
The agencies often **overstated** their **usefulness**
Fontana, NC
40,000 new dams 1950-1980
U.S. RIVERS
Oroville, CA
Feather River
Often required thousands of workers
and **understated** their **harmful** effects.
Lock and Dam, Mississippi

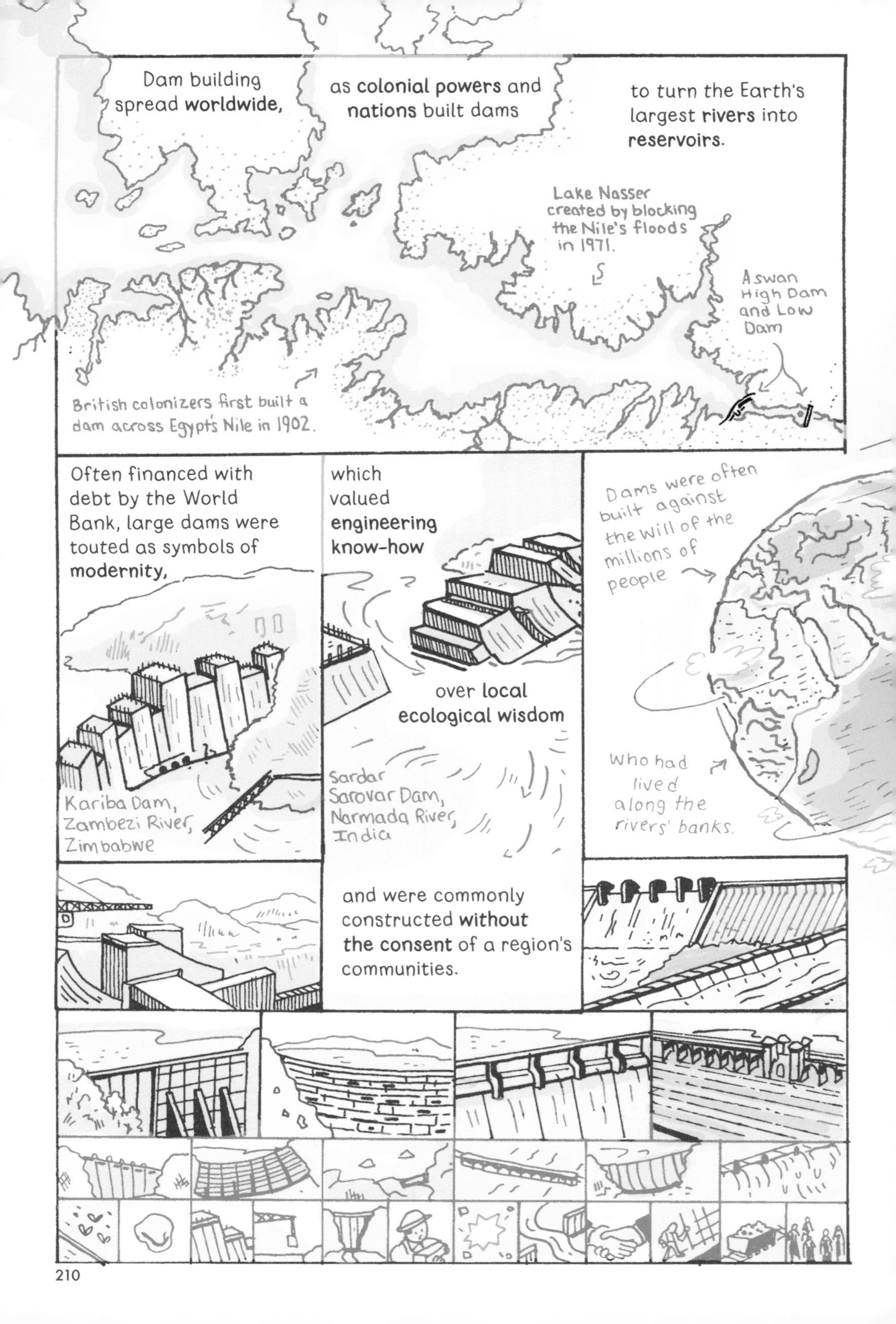
Dam building spread **worldwide,**
as **colonial powers** and **nations** built dams
to turn the Earth's largest **rivers** into **reservoirs.**
Lake Nasser created by blocking the Nile's floods in 1971.
Aswan High Dam and Low Dam
British colonizers first built a dam across Egypt's Nile in 1902.
Often financed with debt by the World Bank, large dams were touted as symbols of **modernity,**
Kariba Dam, Zambezi River, Zimbabwe
which valued **engineering know-how**
over **local ecological wisdom**
Sardar Sarovar Dam, Narmada River, India
Dams were often built against the will of the millions of people
who had lived along the rivers' banks.
and were commonly constructed **without the consent** of a region's communities.

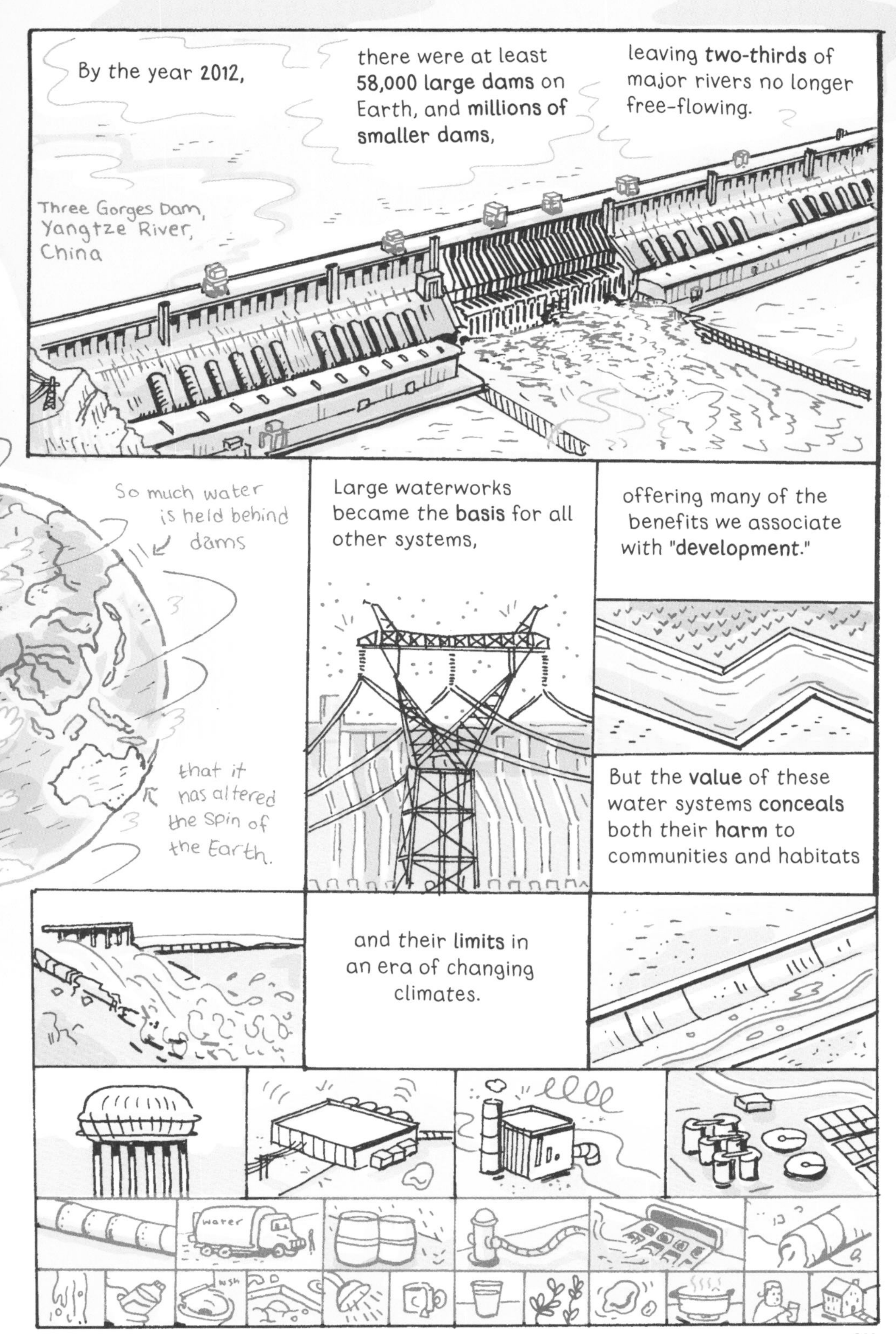
By the year 2012,
there were at least 58,000 large dams on Earth, and millions of smaller dams,
leaving two-thirds of major rivers no longer free-flowing.
Three Gorges Dam, Yangtze River, China
So much water is held behind dams
that it has altered the spin of the Earth.
Large waterworks became the basis for all other systems,
offering many of the benefits we associate with "development."
But the value of these water systems conceals both their harm to communities and habitats
and their limits in an era of changing climates.
water

5. Scarcity and Abundance

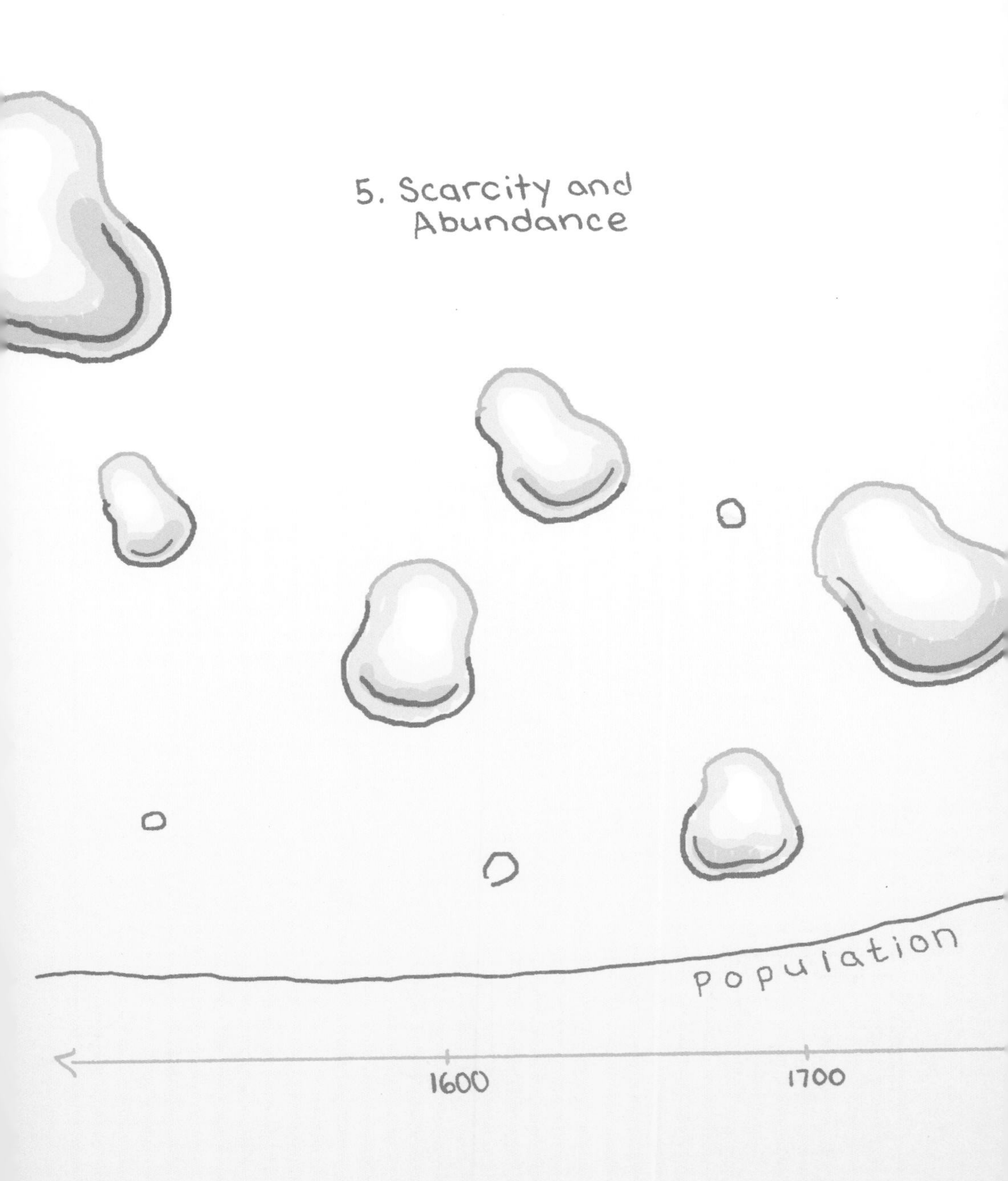

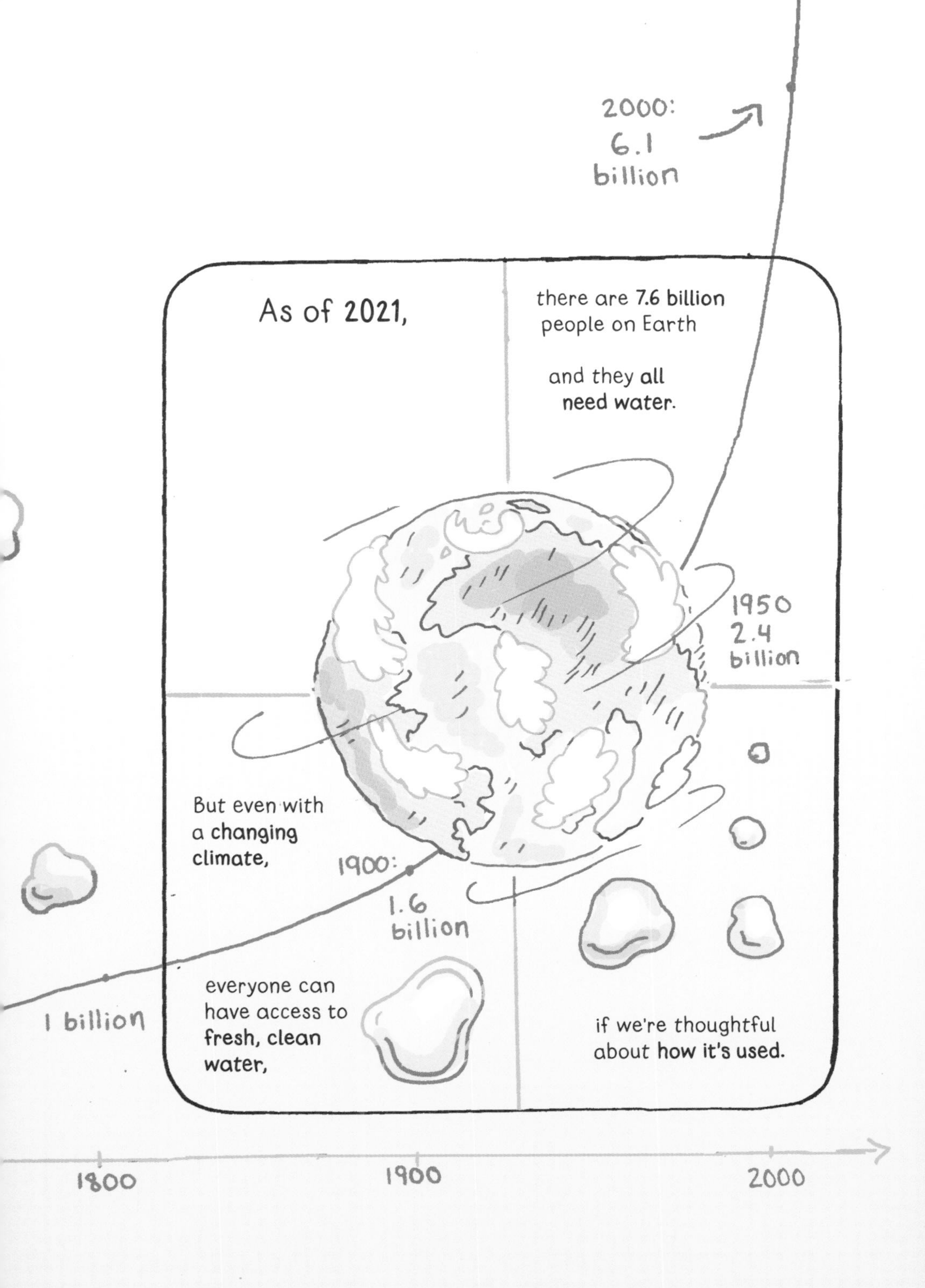
2000:
6.1
billion
As of 2021,
there are 7.6 billion
people on Earth
and they all
need water.
1950
2.4
billion
But even with
a changing
climate,
1900:
1.6
billion
everyone can
have access to
fresh, clean
water,
1 billion
if we're thoughtful
about how it's used.
1800
1900
2000

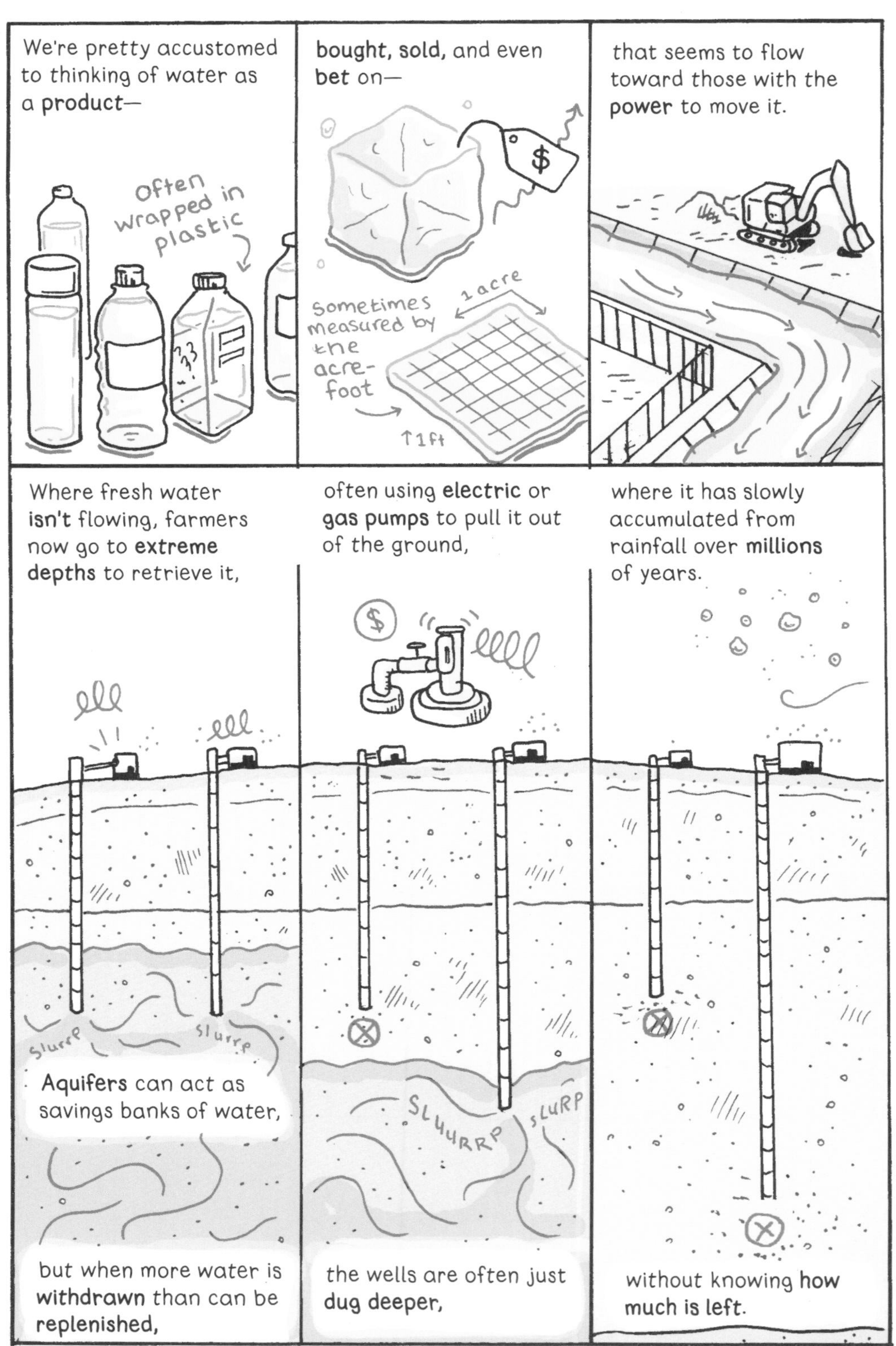
We're pretty accustomed to thinking of water as a **product**—
Often wrapped in plastic
bought, sold, and even **bet** on—
$
sometimes measured by the acre-foot
1 acre
↑1ft
that seems to flow toward those with the **power** to move it.
Where fresh water **isn't** flowing, farmers now go to **extreme depths** to retrieve it,
slurrp
slurrp
Aquifers can act as savings banks of water,
but when more water is **withdrawn** than can be **replenished,**
often using **electric** or **gas pumps** to pull it out of the ground,
$
SLUURRP
SLURP
the wells are often just **dug deeper,**
where it has slowly accumulated from rainfall over **millions** of years.
without knowing **how much is left**.

Water is withdrawn for **agriculture,** which makes sense,
alfalfa
but an inordinate amount is used for **meat-based diets**
used to feed cattle
and to create excesses of **cash crops** like cotton and nuts.
$
Water is held by dams and used for its weight to **turn turbines,**
and used to **transfer heat** in most power plants
to create **electric current.**
It's used by **industry,**
to produce chemicals
and refine oil
to cool data centers,
make clothes,
one pair of jeans requires over 2,000 gallons of water.
~3,000 gallons
600 gallons
and basically **everything else** around us.
30 gallons
~600 gallons

For the little water we withdraw for drinking and personal use,
Pumped up
Weight
some of us have access to brilliant systems of **pipes, pumps,** and **towers,**
Water pressure
providing water that's been **specially filtered** and **treated** for human consumption.
For the billions of people **without** access to water systems,
they still have to collect water daily
by whatever means possible.
WATER
This means contending with the **scarcity,**
industrial pollution,
300-400 million tons of heavy metals, toxic sludge, and other waste is dumped each year into the world's waters.
and **sewage** created by the water use of others.
80% of waste water flows back into the environment untreated.
1.8 billion people drink feces-contaminated water.

In the U.S., even **access** to waterworks doesn't guarantee safety.

When systems are built and then neglected, they can carry pollution that quietly poisons entire populations,

usually found in Black and Brown communities that lack **political** and **economic support.**

can cause neurological harm and more

lead pipes

especially dangerous for Children

Even in wealthy cities, the difficult and costly upgrades can take **decades** and are often put off by elected leaders.

New York has long been building a **two-mile** bypass tunnel **900 feet under** the Hudson River

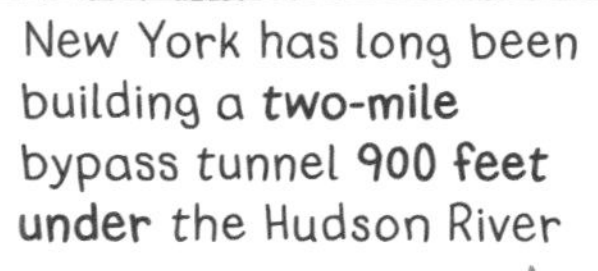

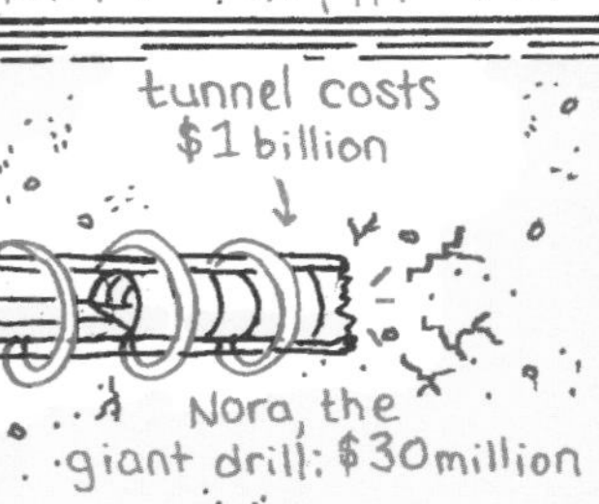

just so it can repair a single leaky aqueduct that supplies more than **half** the city's water.

Waterworks, like all infrastructure, are never really **finished**—

communities need to **keep them up, reimagine,** and **rebuild them,**

especially **now,**

as the cycles they were based on begin to **change rapidly**.

The global industrialization driven by the U.S. and Europe has begun to **change** Earth's cycles,
shifting the long-held patterns that all life has come to rely on
and creating **wild, unexplained outbursts** of weather
that we never planned for.

These changes can bring both **sudden**, violent **surges of water**
and **decades** of a **desolate lack of it,**
potentially forcing **entire populations** to move
to more **viable land**
with more **reliable water cycles.**

We talk a lot about **new technologies** that can help us **ease** some of these issues.
Desalination plants are giant machines that can **remove salt from water,**
but like everything, these come with their own **limitations** and **ecological cost.**
requires LOTS of energy
produces briny waste
We can slow the Earth's warming by **changing our sources of energy,**
carbon in the air
carbon from the ground
using **as little carbon fuel** as possible
to **power** and **heat** homes and move ourselves around.
And we can use the **internet, sensors,** and **computer models**
to understand **patterns** on a global scale,
predict changes, and **inform policy.**
2050

But instead of putting technological solutions first,
hmm
we need to act on an understanding of how water use impacts others—
and of our place within the water cycle.
We need to make decisions that recognize the power disparities between the communities upstream of these cycles
and those downstream,
?
and use that power unselfishly to share water across entire populations.
And we need human solutions that support those displaced by water's shifting cycles,
dropping imaginary pricing and valuations
and arbitrary borders
when it comes to sharing access to a basic human need.

Recently, communities have moved toward using the benefits of **natural infrastructure,**
Urban bioswales soak up stormwater
Urban trees reduce air temperature
stores carbon
absorbs runoff
looks nice
Natural floodplains slow floods and reduce their height
taking down **dams to let rivers flow freely,**
River restoration is often led by Indigenous groups.
Glines Canyon, WA
returning **stolen land** and **water** to local Indigenous communities,
LANDBACK
Rapid City, SD
Lakota Land
and turning toward the wisdom of those who have long cared for it.
Naso Forests, Panama
Teribe River

In each of these actions, we recognize that **water cycles,**
used by **vast ecosystems,**
are what **animates biodiversity**
and its **infinite symbiotic relationships**
at every scale imaginable.

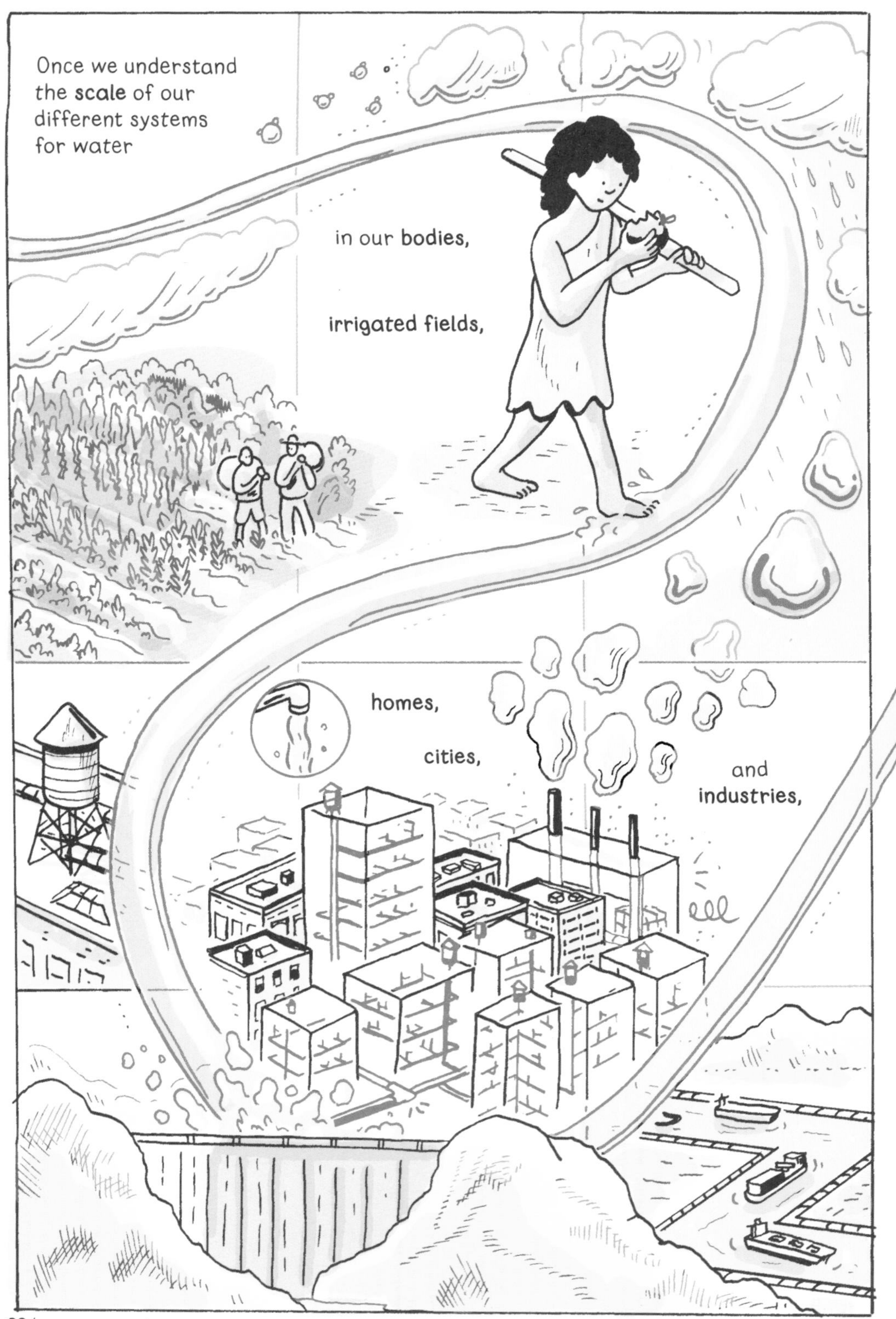

Once we understand the **scale** of our different systems for water
in our **bodies,**
irrigated fields,
homes,
cities,
and **industries,**

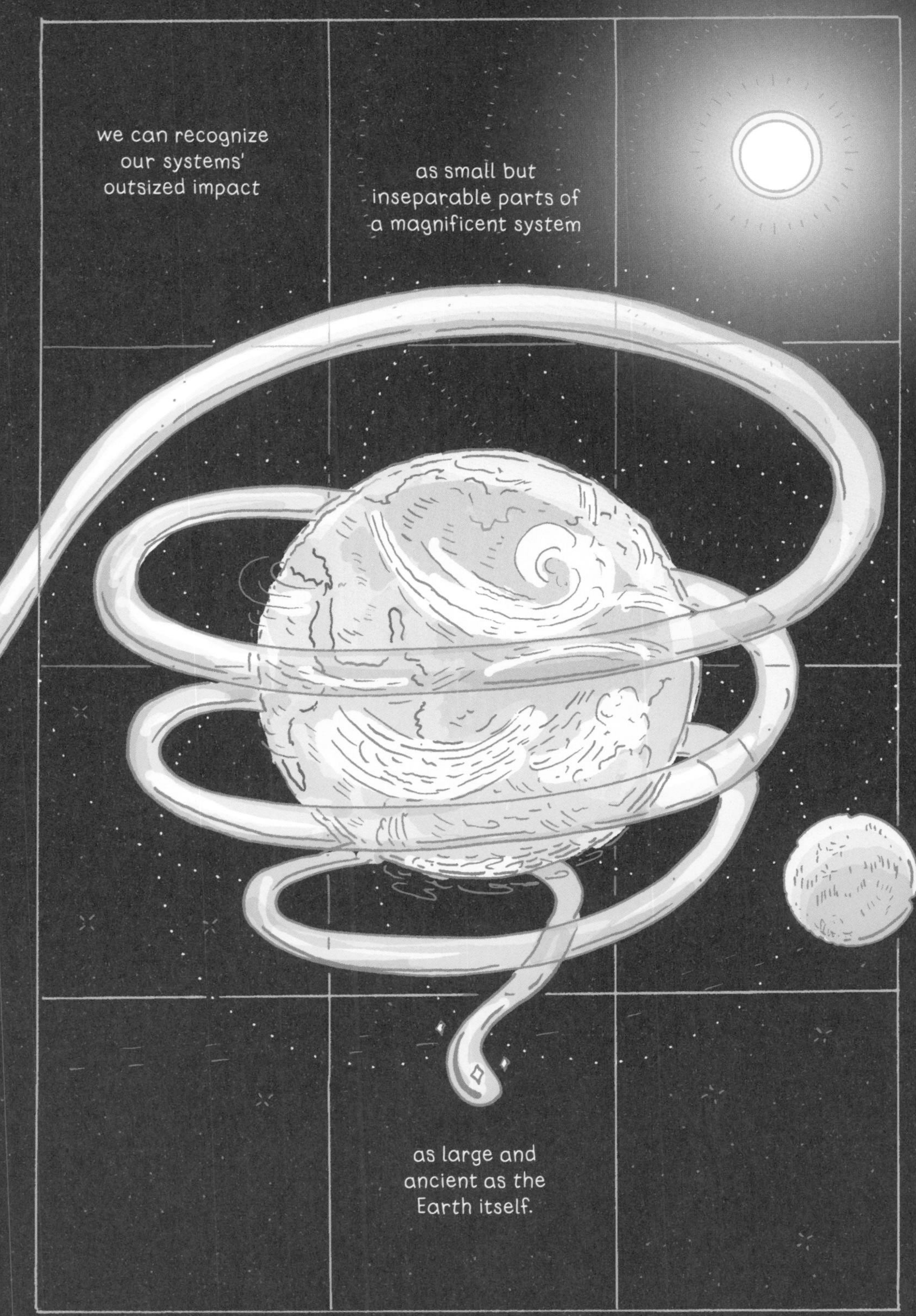
we can recognize
our systems'
outsized impact
as small but
inseparable parts of
a magnificent system
as large and
ancient as the
Earth itself.

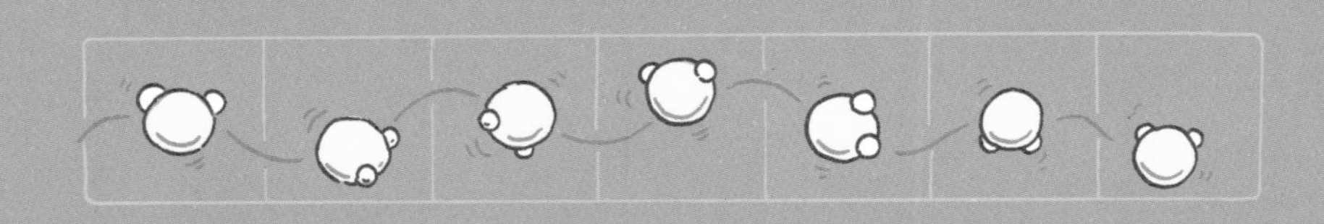

Conclusion:

What are the futures we can imagine?

Even after making this book,
I am very surprised by how much I still don't know.
I realized there's a big difference between the "past,"
which is infinite,
(and kind of overwhelming)
and "histories,"
which are subjective and selective attempts to record and interpret what's happened before.
Any reader who tried to answer the same questions as I did for this book
would probably have chosen a different narrative
to write and draw in all these little boxes.
Like whether to zoom in on little details
or zoom out to convey an understanding from a wider perspective.

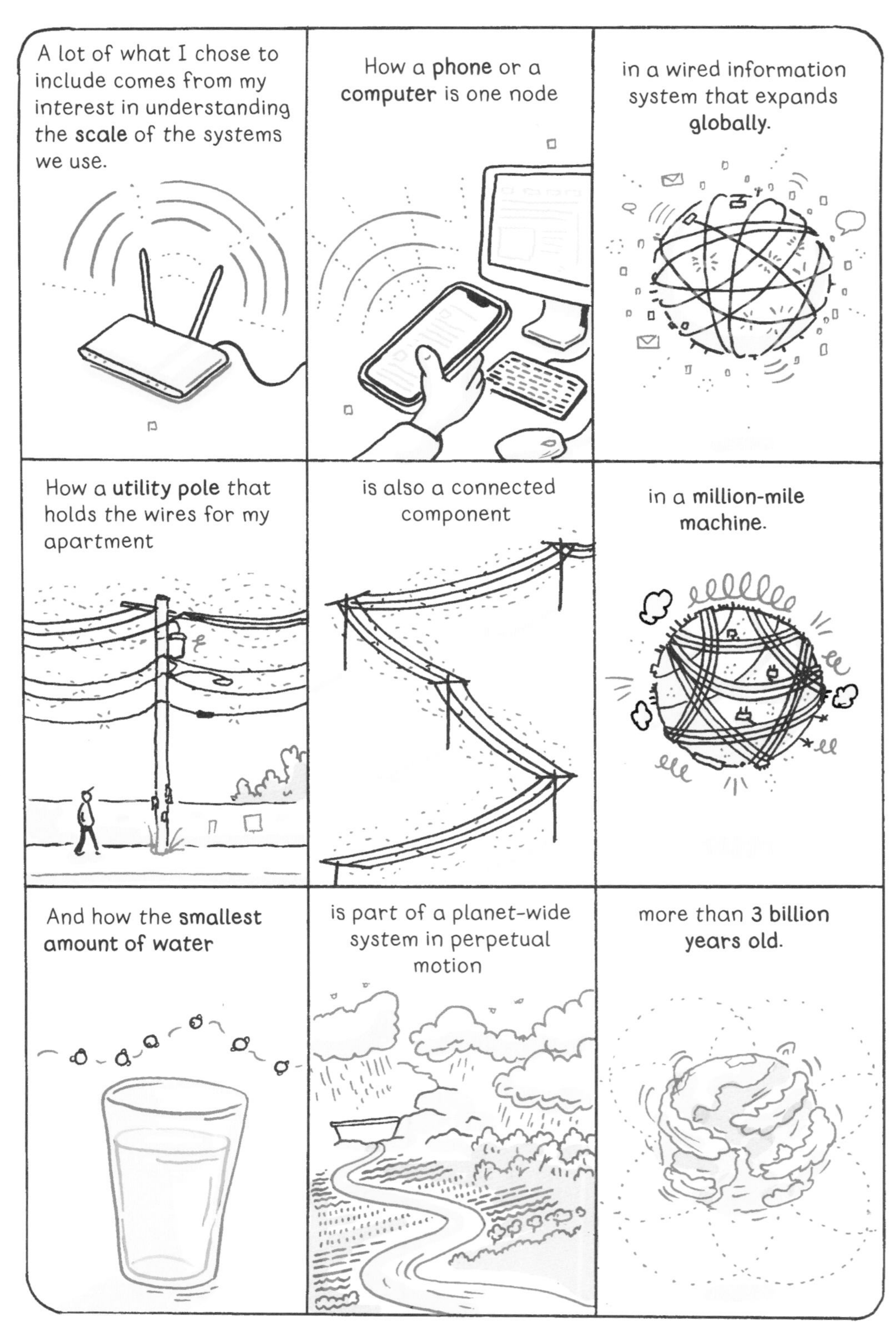
A lot of what I chose to include comes from my interest in understanding the **scale** of the systems we use.
How a **phone** or a **computer** is one node
in a wired information system that expands **globally**.
How a **utility pole** that holds the wires for my apartment
is also a connected component
in a **million-mile machine**.
And how the **smallest amount of water**
is part of a planet-wide system in perpetual motion
more than **3 billion years old**.

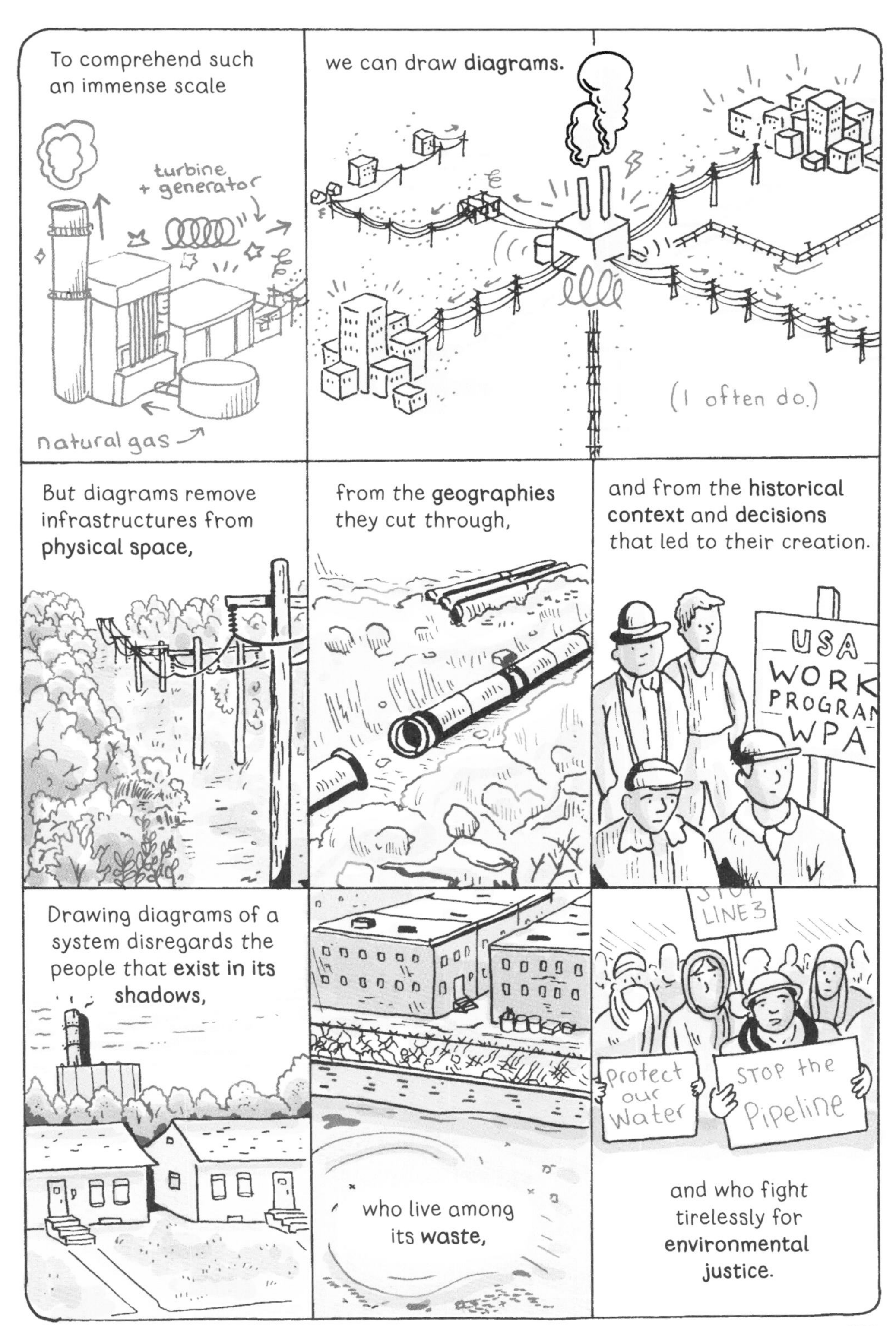
To comprehend such an immense scale
turbine + generator
natural gas
we can draw **diagrams.**
(I often do.)
But diagrams remove infrastructures from **physical space,**
from the **geographies** they cut through,
and from the **historical context** and **decisions** that led to their creation.
USA
WORK
PROGRA
WPA
Drawing diagrams of a system disregards the people that **exist in its shadows,**
who live among its **waste,**
LINE 3
Protect our Water
STOP the Pipeline
and who fight tirelessly for **environmental justice.**

Hidden systems are so much **more than engineering**—
they shape how we **live** and **think.**
I try to imagine the first time seeing
a **light** that didn't produce **smoke,**
or the **beauty** of a
freshwater fountain,
having only ever
drunk **dirty,**
brackish water.
Or witnessing the first
message sent instantly
across an ocean via a
telegraph cable
TELETYPE
or through a
computer network.

All of these systems were once **marvels,**
and now make possible everything we do.
But each comes with a **cost**
flush
to **others,**
and to the **earth** itself.

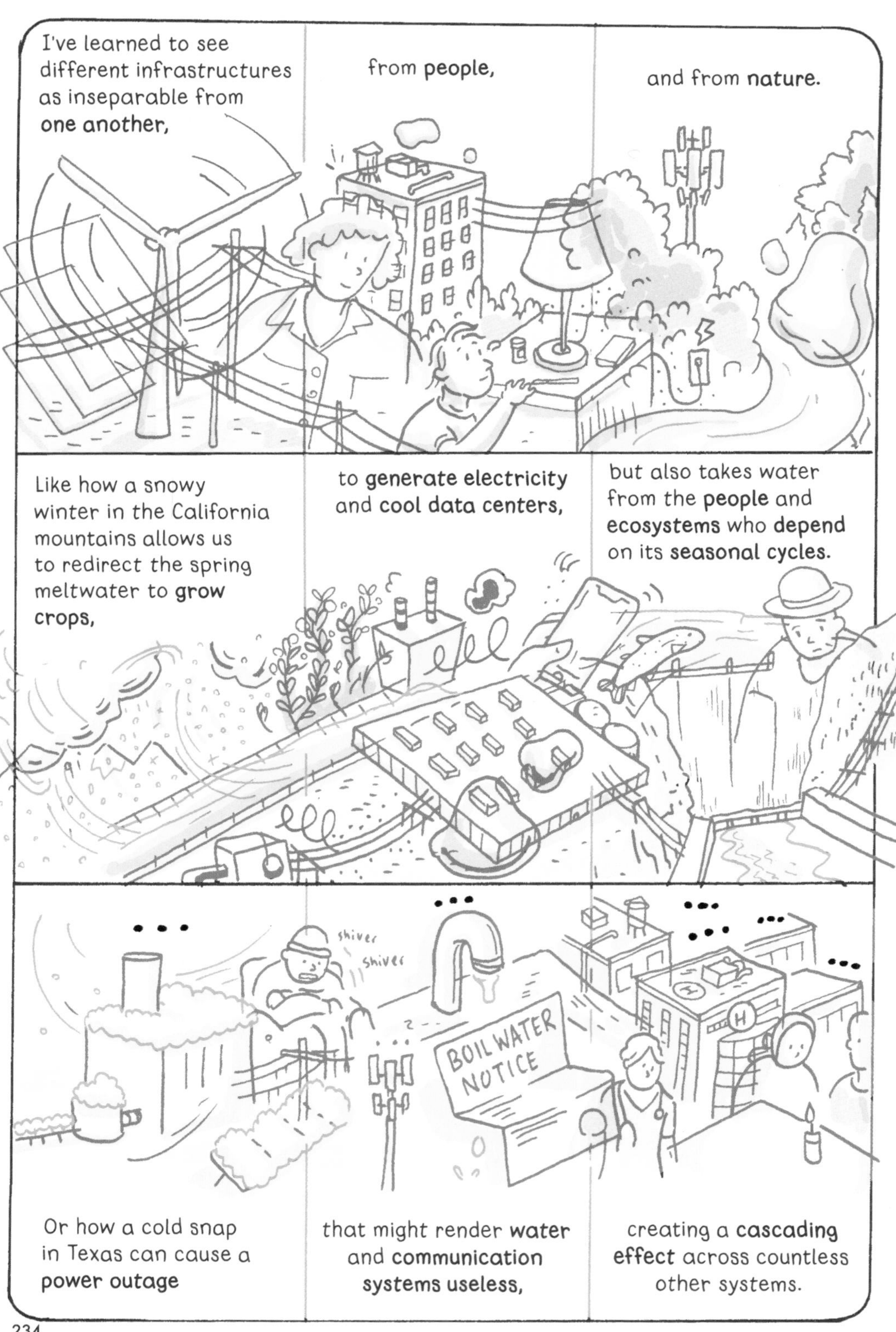
I've learned to see different infrastructures as inseparable from **one another,**
from **people,**
and from **nature.**
Like how a snowy winter in the California mountains allows us to redirect the spring meltwater to **grow crops,**
to **generate electricity** and **cool data centers,**
but also takes water from the **people** and **ecosystems** who **depend** on its **seasonal cycles.**
shiver
shiver
BOIL WATER NOTICE
Or how a cold snap in Texas can cause a **power outage**
that might render **water** and **communication systems useless,**
creating a **cascading effect** across countless other systems.

And while information, electricity, and water often move **invisibly,**
understanding these systems while they're **working**
can be its **own way of seeing**
and of appreciating the **natural foundations** these systems rest on.

People initially built infrastructure by following the contours of Earth's natural systems,

and new systems tended to **follow** in the paths previously plotted.

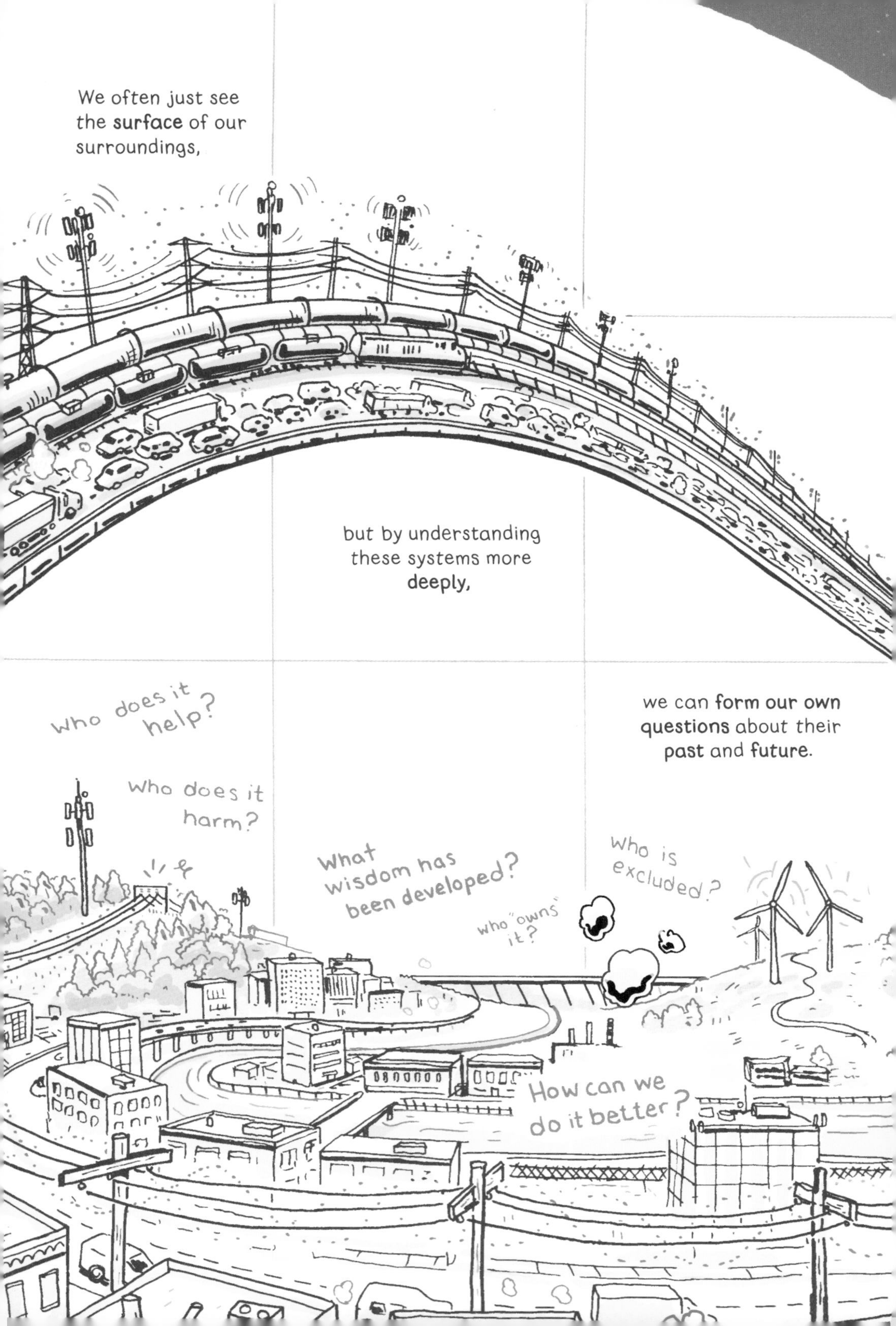
We often just see the **surface** of our surroundings,
but by understanding these systems more **deeply,**
we can **form our own questions** about their **past** and **future.**
Who does it help?
Who does it harm?
What wisdom has been developed?
Who "owns" it?
Who is excluded?
How can we do it better?

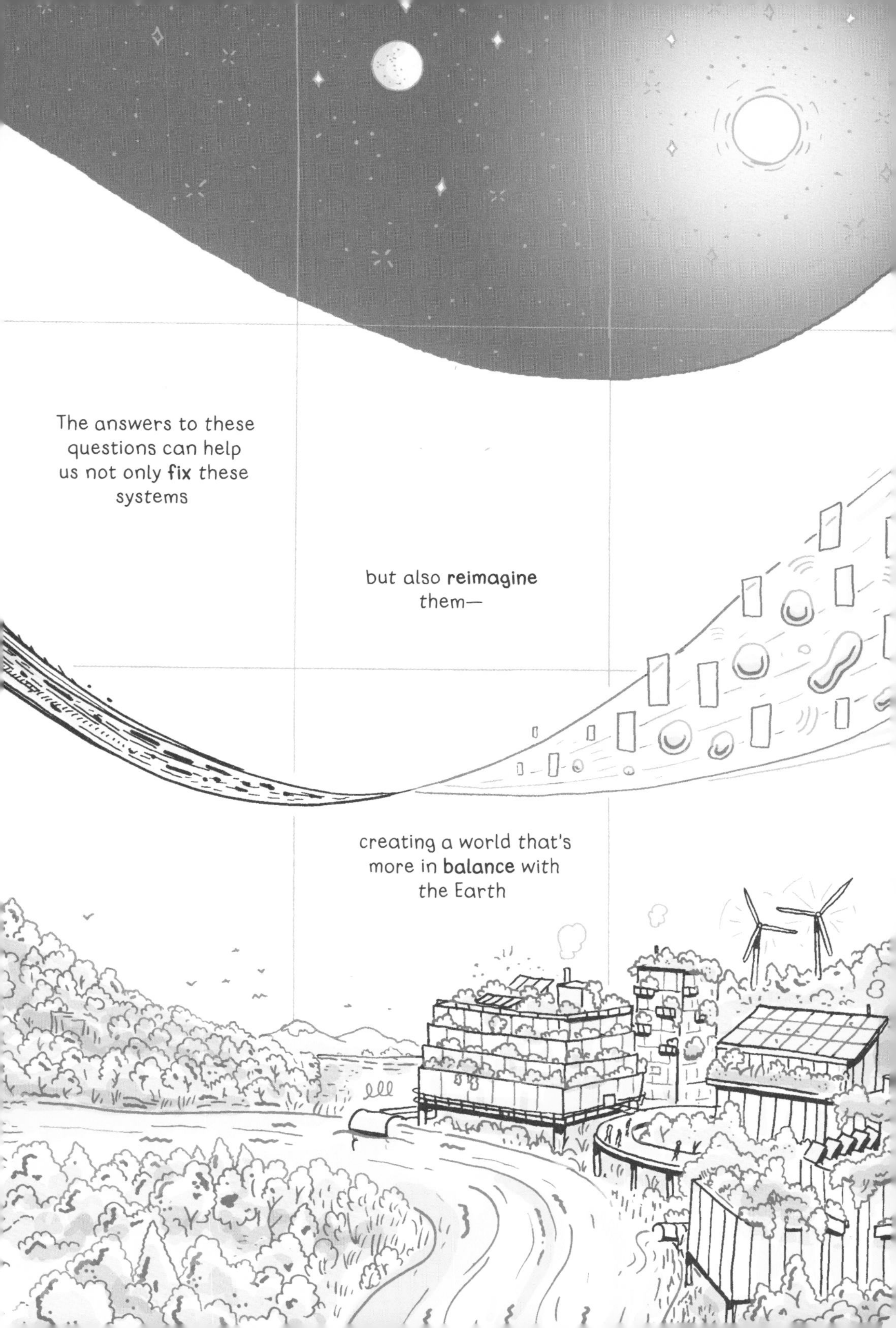
The answers to these questions can help us not only **fix** these systems
but also **reimagine** them—
creating a world that's more in **balance** with the Earth

and that **provides** equitably for **all people**.

bit of information
electric current
spinning, electrical generation
transforming current
no power
wireless infor-mation.
pollution,
Hidden Systems Sketchbook
48-Page Memo Book
Materials / Made in the U.S.A.
period

We picture...
FABER-CASTELL
Pitt Artist Pen
TINTA DE ARCHIVE
0.35mm line / ligne / lines
traffic passed.

Units of measurement

Data	megabyte unit of data MB a song or photo might be ~5 MB	gigabyte 1,000 megabytes GB movie ~10 GB	terabyte 1,000 gigabytes TB ~200,000 pictures
power	Watt unit of power W LED light bulb ~10 W	Kilowatt 1,000 watts KW VRRR AC appliances might use ~1-2 KW	Megawatt 1,000 Kilowatts MW power plants might produce 50-2,000 MW
Water	Gallon Unit of volume gal 1 gal jug 3.7 Liters	Cubic foot ~7.5 gallons ft^3 a bathtub might use 5	acrefoot ~326,000 gal ac.ft 1 acre, 1 foot deep

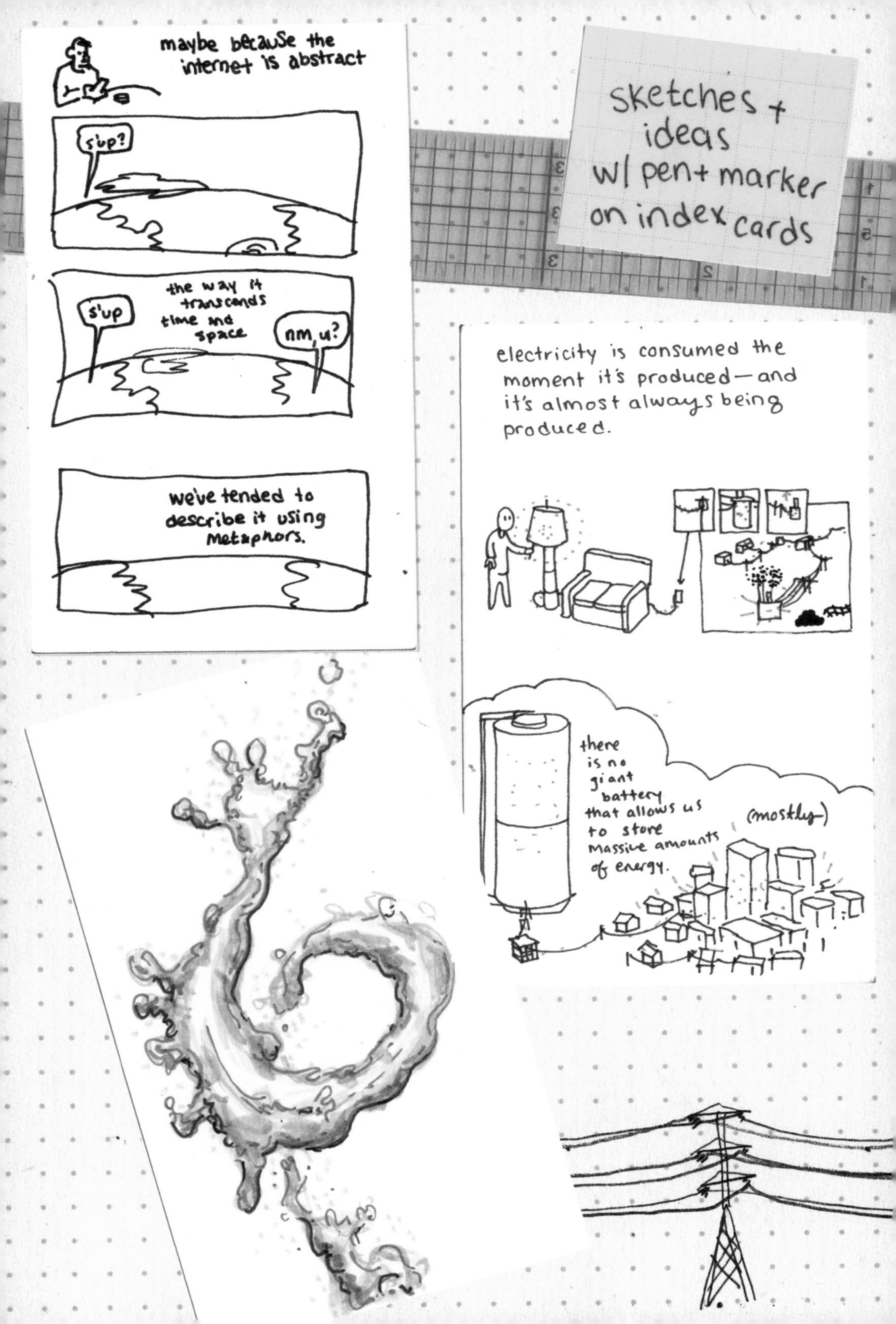
maybe because the internet is abstract
s'up?
the way it transcends time and space
s'up
nm, u?
we've tended to describe it using metaphors.
sketches + ideas w/ pen+ marker on index cards
electricity is consumed the moment it's produced—and it's almost always being produced.
there is no giant battery that allows us to store massive amounts of energy.
(mostly)

old sunken utility poles along train tracks next to the CT River

Studio in an
old telegraph
building

secret internet building,
white River Junction, Vt
822 822

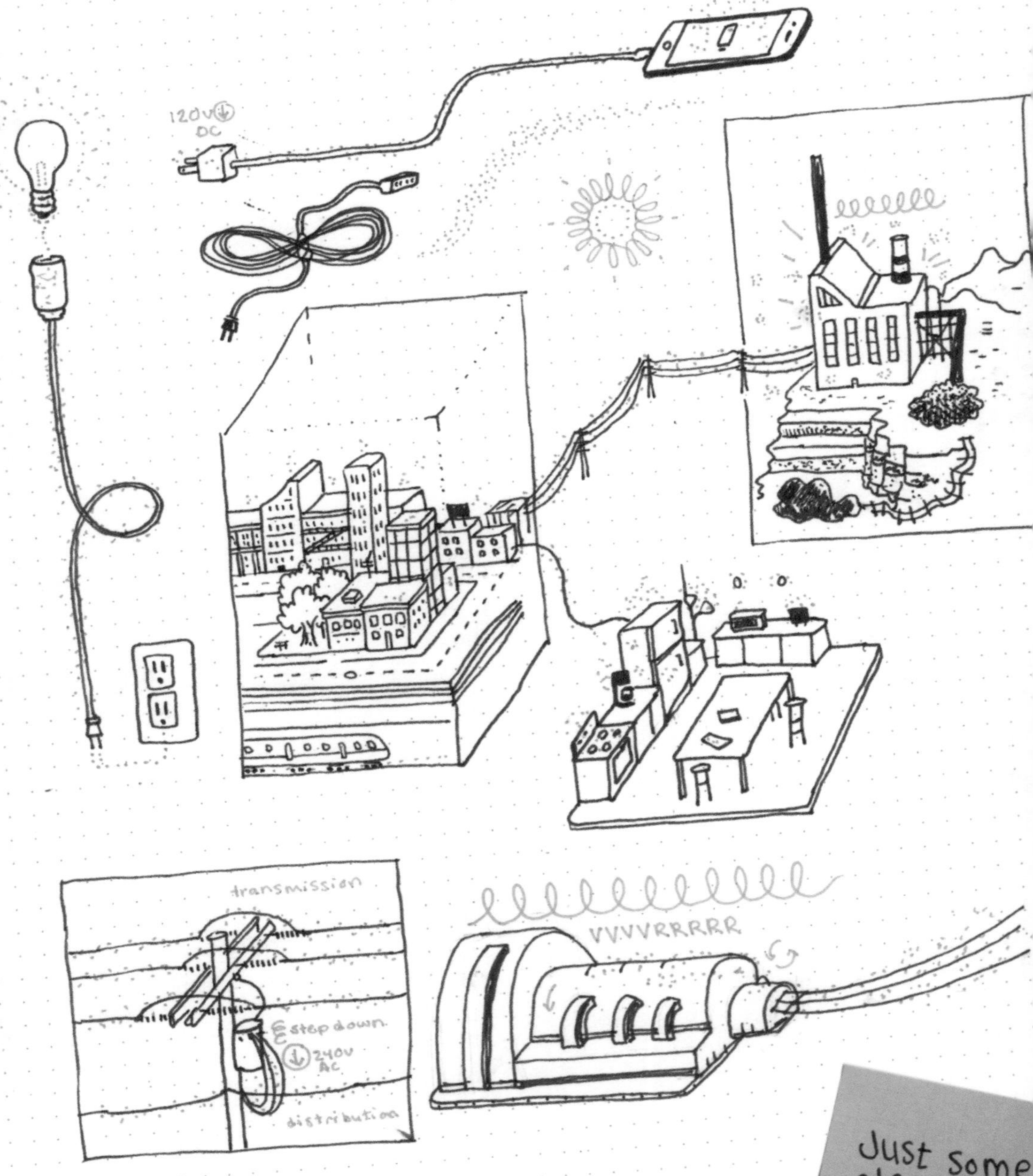
120V
DC
transmission
step down
240V
AC
distribution
VVVVRRRRR
Just some
electricity
doodles

Substation
in Wilder, VT
DANGER
KEEP OUT

Author's Notes and Acknowledgments

Time Frames: These comics were researched, written, and drawn one after the other over the course of four years. Every day I hear about new developments—additional pieces being added, new policies, victories, and setbacks in the struggle to make them more equal for everyone. I've updated much of the information during the process, but please consider the time frame in which these stories were researched and drawn to ask questions about what topics to stay updated on.

Lines of Light: 2017-2018
Power Grid: 2018-2019
Water Cycle: 2020-2021

Land Acknowledgment: This book was created in Ndakinna (meaning "our land"), the traditional unceded homeland of the Western Abenaki, at the junction of Wôbitekw (White River) and Kwanitekw (Connecticut River) in what is now known as the Upper Valley of Vermont and New Hampshire. Abenaki people continue to live across Ndakinna, which includes areas in Vermont, New Hampshire, northern Massachusetts, western Maine, and southern Quebec.

The Team That Made This Book: Thank you, Daryl, for your love and support and for teaching me so much about comics. Thank you to my family—Mom, Dad, David, Lee, and Ed—and all the friends who've asked me over the years how my book is going. Thank you to the faculty and community at the Center for Cartoon Studies.

Thank you, Gina Gagliano, for your wonderful insight, enthusiasm, and patient support in helping me develop and edit this book, and thank you to Whitney and Patrick at RH Graphic for your amazing work sending it out into the world. To my agent, Farley Chase, for seeing a future for these stories. To Andy Warner, who provided valuable guidance when this book began as an MFA thesis project at the Center for Cartoon Studies.

Feedback and Support: Thank you to those who read drafts and provided feedback and support: Sophie Yanow, Jason Lutes, Daryl Seitchik, Cuyler Hedlund, Leise Hook, Emma Hunsinger, Tillie Walden, Jarad Greene, Issy Manley, Nur Schuba, James Sturm, Meredith Angwin, and Brian Hayes.

Simplifying and Reducing: I made these comics to convey an understanding of the systems but want to acknowledge that the narrative often speeds

through details and history, and that when drawing infrastructure and systems, I'm often prioritizing clarity over visual precision.

Research: The ideas in these comics were influenced by countless sources. The comic draws on the work of people who have traveled to, researched, photographed, and written about the internet, the electric grids, climate, and water systems across the world. To gain a deeper understanding, check out the notes and selected bibliography for some of the books, movies, and art that informed the drawings and ideas in this book.

Thank you to the experts who helped with big and small favors and advice: Meredith Angwin for valuable insight and infrastructure adventures; Molly and Eric at New England ISO; the staff at Green Mountain Power and Northern Kingdom Community Wind Farm, Christine Hallquist, Brian Hayes, Matthew Wald; Candace Clement at Free Press; my high school science teacher, Linda Tarantino; Rick Kenney from the Hartford, Vermont, water department; Mark Wood and Lance Swenson at Consolidated Communications; Francis J. Magilligan, Christopher S. Sneddon, and Coleen A. Fox from the Dartmouth College geography department; Bruce R. James; Eric Sanderson; the amazing librarians at Kilton Public Library in Lebanon, New Hampshire; and the countless individuals who entertained conversations with me about their understanding and relationship to the internet, electricity, and water.

Artistic Influences: It would be impossible (and not very useful) for me to try to tally all the influences that went into this book, but a few stand out for keeping me inspired. The brilliant picture books of Wanda Gág, Virginia Lee Burton, and David Macaulay; the narrative visualizations of time in the comics of Richard McGuire, Kevin Huizenga, and Sophia Foster-Dimino; the comics essays of Sophie Yanow and Sam Wallman; the stunning portrayal of invisible energies in the work of Ron Regé, Jr. and Lale Westvind; and, of course, *Avatar: The Last Airbender*.

". . . the ultimate, hidden truth . . ." quoted from David Graeber's *The Utopia of Rules: On Technology, Stupidity, and the Secret Joys of Bureaucracy,* published by Melville House in 2015.

Lines of Light

"Somehow I knew the notional space . . ." Quoted from "William Gibson: Beyond Cyberspace" by Thomas Jones, *The Guardian,* Sept. 22, 2011.

"Cyberspace. A consensual hallucination . . ." Quoted from *Neuromancer* by William Gibson, published in 1984 by Ace Books.

"The metaphors we use often reflect our bias . . ." I began sketching about this concept over a decade ago while drawing political cartoons—a medium that relies heavily on visual metaphors. I drew upon Josh Dzieza's short article "A History of Metaphors for the Internet," which provided a helpful chronology for some of these terms and further details on the people who have studied and written about them, including Al Gore, Judith Donath, Cornelius Puschmann, Jean Burgess, Peter Lyman, Tim Wu, and Rebecca Rosen.

"The Internet is the mostly physical infrastructure . . ." My understanding of the internet as a physical thing was influenced by some excellent books and writings, particularly Andrew Blum's *Tubes: A Journey to the Center of the Internet;* the work of Ingrid Burrington, including her field guide *Networks of New York: An Illustrated Field Guide to Urban Internet Infrastructure;* and Nicole Starosielski's *The Undersea Network,* which details the implications of the network of submarine cables.

Chapter 2: Cables The boat referenced for this picture was most likely used primarily for laying electrical cables, which are also spooled on ships and laid along the ocean floor. Often, the communication cable ships are bigger with multiple bays for spooling cable within the ship itself.

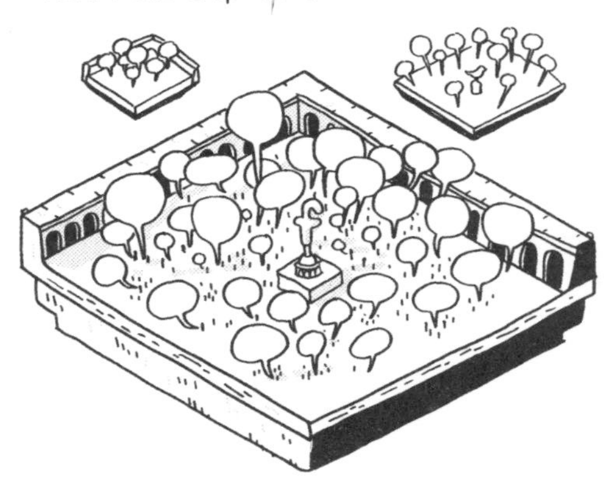

"The All Red Line . . ." map approximated with reference from *The All Red Line: The Annals and Aims of the Pacific Cable Project* by George Johnson, 1903. This map shows only the British-operated cables. There were plenty of other cables around this time, which are represented on the map on page 27.

"When the U.S. attacked Cuba . . ." As far as I can tell, this was the first instance of undersea cable sabotage being used in warfare. There are conflicting accounts about how successful this was—though most agree the U.S. did not manage to cut all the cables.

"Roosevelt used the Pacific Cable . . ." "The Pacific Cable, Hawai'i, and Global Communication" by Jeffrey K. Lyons.

"The telegraph network, through . . ." Map approximated with reference from historical cable company maps. There's a selection of these maps archived by Bill Burns at Atlantic-cable.com. Nicole Starosielski's *The Undersea Network* provides a vivid examination of the colonial history of the telegraph system.

"Today, fiber-optic cables . . ." Maps approximated with reference from TeleGeography's Cable map, which shows imprecise paths of the cables for diagrammatic purposes. Dozens of cables have been put into service from the time I began researching this book to its publication. I updated some—but for an up-to-date interactive map, refer to submarinecablemap.com. The number of cables and total length are based on numbers from TeleGeography's Submarine Cable FAQ, as of late 2021.

"The internet, which initially used . . ." Map of the U.S. fiber backbone approximated from "InterTubes: A Study of the US Long-haul Fiber-optic Infrastructure" by Ramakrishnan Durairajan, Paul Barford, Joel Sommers, and Walter Willinger, for the University of Wisconsin in 2015. Because no comprehensive record of these cables existed, these researchers had to spend years piecing together a map. For a deeper dive into the pathways of railroads and fiber-optic cables, see Ingrid Burrington's talk "It Tends to Annihilate Distance," presented at the Eyeo Festival in 2015.

"By the military to help manage the surveillance of Americans . . ." For a thorough accounting of this history, see *Surveillance Valley: The Secret Military History of the Internet* by Yasha Levine.

"Today, exchange points are . . ." Map approximated with reference from TeleGeography's Internet Exchange map. For a more precise and interactive map, refer to internetexchangemap.com.

"At the local level too . . ." Pictures referenced from an obscure and amazingly comprehensive directory of these unique buildings available at co-buildings.com.

"They're the size of warehouses . . ." They're actually a lot larger—I referenced a particular data center complex based on Google Maps, and I also referenced some photos Ingrid Burrington took for her pieces in the excellent series for the *Atlantic,* "Beneath the Cloud," which ran from 2015 to 2016.

"Data centers are located around the country . . ." This is a made-up data center for diagrammatic purposes.

"Experimenting with balloons . . ." Both Facebook and Google were working on this, but so far balloons and drones haven't been found to be as practical, while satellites appear to hold more promise for providing internet access.

"The Internet looks a lot like a computer . . ." This metaphor of the "internet as a computer" has been explored for decades. In 1984, John Gage said, "The network is the computer." In 1996, Neal Stephenson wrote a long-form essay on the world's longest fiber-optic cable for *Wired* magazine titled "Mother Earth Mother Board," where he wrote: "If the network is The Computer, then its motherboard is the crust of Planet Earth."

Electric Grid

"The farthest Thunder that I heard . . ." This version is from *The Poems of Emily Dickinson,* published by Belknap Press of Harvard University Press in 1999, and edited by R. W. Franklin.

Experiments—For a very comprehensive study of all the strange and sometimes silly early electrical instruments, see Michael Brian Schiffer's *Draw the Lightning Down: Benjamin Franklin and Electrical Technology in the Age of Enlightenment.*

The Pearl Street Station—The drawing of the Pearl Street Station was approximated from various etchings and photographs of a scale model that was produced by the Edison Company in 1927, now housed in the National Museum of American History.

The War of the Currents—The "war of the currents" is perhaps the most famous story to emerge from the history of the electric grid, and there are dozens of books and movies that detail the history. For an accounting of Edison's work on the electric chair, see Mark Essig's *Edison and the Electric Chair: A Story of Light and Death.*

Building the Grid—Gretchen Bakke provides details of the tangle of wires blocking out the sky in New York in her excellent and accessible history *The Grid: The Fraying Wires Between Americans and Our Energy Future.* For a pioneering look at the social and cultural impact of the electric grids, such as their role in enabling suburbs, see David E. Nye's *Electrifying America: Social Meanings of a New Technology, 1880-1940.*

The New Deal—The top panel references the precisionist painter Charles Sheeler's *Suspended Power,* made in 1939, in the collection of the Dallas Museum of Art.

World War II—Julie A. Cohn describes how holding companies and World War II affected the interconnection of the power grids in her book *The Grid: Biography of an American Technology.* Interconnection was not a linear process, and some utilities disconnected following the wartime government mandate.

After the War—This page references an episode of *General Electric Theater,* which aired in 1956 and included a look at the electric household of future president Ronald Reagan. According to General Electric, in 1956, it was the third-most-popular show on television, reaching over 25 million viewers a week.

Growth of the Grid—These maps are approximated from figures featured in "Report on the Status of Interconnected Power Systems," published by the Edison Electric Institute in 1962, and were featured in Julie A. Cohn's *The Grid: Biography of an American Technology.*

Coal-Burning Plants—For excellent and creative research into the role of coal and electricity in development of the cities of the North American Southwest, see Andrew Needham's *Power Lines: Phoenix and the Making of the Modern Southwest.*

Power Mix—Visualizing a local power mix is a difficult task. To see how your state produces power, see "How Does Your State Make Electricity?" by Nadja Popovich and Brad Plumer.

Balancing Authorities—This drawing is based on the control center for the New England Independent System Operator (NE-ISO), which balances the electricity on the New England grid. The regulations and markets involved in electricity generation and transmission are impossibly complex. Meredith Angwin discusses these hidden systems in *Shorting the Grid: The Hidden Fragility of Our Electric Grid.*

Around the World—For more on the colonial politics of Zimbabwe's electric grid, see Moses Chikowero's paper "Subalternating Currents: Electrification and Power Politics in Bulawayo, Colonial Zimbabwe, 1894-1939." For information on Puerto Rico's struggles with their electric grid, see "Privatizing Puerto Rico," by Ed Morales.

Waterworks

"You know, they straightened out the Mississippi River . . ." This quote, excerpted from Toni Morrison's essay "The Site of Memory," continues: "Writers are like that: remembering where we were, what valley we ran through, what the banks were like, the light that was there and the route back to our original place. It is emotional memory—what the nerves and the skin remember as well as how it appeared. And a rush of imagination is our 'flooding.' "

"Earth formed around 4 billion . . ." For engaging visualizations of Earth's history, see Virginia Lee Burton's picture book *Life Story,* and the Smithsonian's web application "Travel Through Deep Time With This Interactive Earth."

"Relative to its size . . ." Earth water data courtesy of the U.S. Geological Survey (USGS).

"It's easy for us to visualize . . ." Estimates of daily water use vary widely, but 80 to 100 gallons per person per day in the U.S. is an estimate from USGS.

"It's equally hard for us to . . ." The National Science Foundation runs the Ice Core Facility and has some great resources available at icecores.org. Scientists also study tree rings to gain insight and corroborate data about the past.

"Around 6,000 years ago . . ." For wide-ranging histories of water and civilization, see Steven Solomon's *Water: The Epic Struggle for Wealth, Power, and Civilization* and Steven Mithen's *Thirst: Water and Power in the Ancient World.*

"An estimated population of around 1 million people . . . " Ancient population estimates almost always vary widely, and I found estimates ranging from less than half a million to over a million while researching Rome's population.

Civilizational Cycles—Here's my attempt to reference some ingenious waterworks throughout history that I wished I had more space for (and many more are left out still). Victoria Lautman has a beautiful collection of Indian stepwell photography called *The Vanishing Stepwells of India.* And the Virtual Angkor Project allows visitors to explore the 13th-century hydraulic metropolis of Angkor Wat in present-day Cambodia.

Water for Drinking—See Welikia.org for more information on the Mannahatta and Welikia Projects, which aimed to re-create the island of Manhattan and surrounding areas before colonization. The project was led by Eric W. Sanderson.

Croton Aqueduct—Compliments to the Friends of the Old Croton Aqueduct, which has collected and digitized many images showing the construction and celebration of New York's early water systems.

Water in the City—For a full accounting of sewer systems, see *An Underground Guide to Sewers*, by Stephen Halliday, which contains a wealth of visuals on the development of sewer systems across time and geography.

Water for Work—David Macaulay's *Mill* provides an excellent visual guide to how different mills were constructed, from small waterwheels to factories.

Water for Food—The bottom panels on these two pages are inspired by some sequences from Terrence Malick's *Days of Heaven* (1978), set in 1916 in the Texas Panhandle. The film was shot by Néstor Almendros and Haskell Wexler. For more on the environmental disaster of our own making, see *The Dust Bowl* by Ken Burns, which aired on PBS in 2012. For an accounting of the environmental injustice on Native American nations from colonization of North America, see Dina Gilio-Whitaker's *As Long as Grass Grows.*

Water for Irrigation—The Central Valley Project continued for decades and was accompanied by the California State Water Project. The map is exaggerated and approximated from an image commissioned by the United States Bureau of Reclamation, drawn by A. A. Abel and printed by A. Hoen and Co.

Shasta Dam—Some of the images in here are inspired by a silent film called *So Shasta Dam Was Built,* filmed by Howard Colby in 1945 and posted to YouTube by the Shasta Historical Society in 2017.

"Where fresh water isn't flowing . . ." For decades, the pumping of groundwater has been largely unregulated, and the overpumping of aquifers has in some cases literally caused the ground to sink.

"Of the little water we use . . ." This chapter is very light on modern water treatment systems. For a visual understanding of contemporary waterworks and every other type of infrastructure, I highly recommend Brian Hayes's *Infrastructure: A Guide to the Industrial Landscape.*

"Potentially forcing entire populations . . ." The effects of climate change on migration are reported on by Todd Miller in *Storming the Wall: Climate Change, Migration and Homeland Security.*

"Taking down dams to let rivers flow freely . . ." Indigenous groups and tribes are frequently involved in dam removal, and the process is often about cultural healing as well as ecosystem restoration.

For more, see the paper by Coleen A. Fox and others: "'The River Is Us; The River Is in Our Veins': Re-defining river restoration in three Indigenous communities."

"Returning stolen land . . ." This panel references an action taken by activists from ten different Nations on July 4, 2021. For more information, see the NDN Collective at NDNcollective.org.

"Who had long cared for it . . ." For more information on the Naso's efforts to protect their ancestral homeland, see "We Are Nature's Best Guardians, Not the State" by Gabriella Rutherford for intercontinentalcry.org, August 20, 2019.

Conclusion

"And who fight tirelessly for environmental justice . . ." Dina Gilio-Whitaker provides a history of Indigenous resistance to environmental injustice in *As Long as Grass Grows.*

Selected Bibliography

Lines of Light

Blum, Andrew. *Tubes: A Journey to the Center of the Internet.* New York: HarperCollins, 2012.

Burrington, Ingrid, and Emily Ann Epstein, Tim Hwang, Karen Levy, and Alexis Madrigal. "Beneath the Cloud" series, *The Atlantic,* 2015–16.

Burrington, Ingrid. *Networks of New York: An Illustrated Guide to Urban Internet Infrastructure.* Brooklyn: Melville House, 2016.

Ceruzzi, Paul E. *Internet Alley: High Technology in Tyson's Corner, 1945–2005.* Cambridge: The MIT Press, 2011.

Dzieza, Josh. "A History of Metaphors for the Internet." TheVerge.com, 2014. theverge.com/2014/8/20/6046003/a-history-of-metaphors-for-the-internet.

Hayes, Brian. "The Infrastructure of the Information Infrastructure." *American Scientist* 85, 1997.

Hu, Tung-hui. *A Pre-History of the Cloud.* Cambridge: The MIT Press, 2015.

Johnson, George. *The All Red Line: The Annals and Aims of the Pacific Cable Project.* Ottawa: James Hope and Sons, 1903.

Lee, Timothy B. "40 Maps That Explain the Internet." Vox.com, 2014. vox.com/a/internet-maps.

Leiner, Barry M., Vinton G. Cerf, David D. Clark, Robert E. Kahn, Leonard Kleinrock, Daniel C. Lynch, Jon Postel, Larry G. Roberts, and Stephen Wolff. *A Brief History of the Internet.* Internetsociety.org, 1997. internetsociety.org/internet/history-internet/brief-history-internet.

Levine, Yasha. *Surveillance Valley: The Secret Military History of the Internet.* New York: Public Affairs, 2018.

Lyons, Jeffrey K. "The Pacific Cable, Hawai'i, and Global Communication." *The Hawaiian Journal of History* 39, 2005.

Mendelsohn, Ben. *Bundled, Buried, and Behind Closed Doors.* 2011. Video, 10:05. vimeo.com/30642376.

Parker, Matt, dir. *The People's Cloud.* 2017. thepeoplescloud.org.

Rosen, Rebecca J. "Clouds: The Most Useful Metaphor of All Time?" *The Atlantic,* Sept. 30, 2011. theatlantic.com/technology/archive/2011/09/clouds-the-most-useful-metaphor-of-all-time/245851.

Starosielski, Nicole. *The Undersea Network.* Durham: Duke University Press, 2015.

Stephenson, Neal. "Mother Earth Mother Board." *Wired,* 1996.

TeleGeography. Submarine Cable Frequently Asked Questions. Submarine Cable 101. 2021. www2.telegeography.com/submarine-cable-faqs-frequently-asked-questions.

Power Grid

Angwin, Meredith. *Shorting the Grid: The Hidden Fragility of Our Electric Grid.* Hartford: Carnot Communications, 2020.

Bakke, Gretchen. *The Grid: The Fraying Wires Between Americans and Our Energy Future.* New York: Bloomsbury, 2016.

Bodanis, David. *Electric Universe: The Shocking True Story of Electricity.* New York: Crown, 2005.

Chikowero, Moses. "Subalternating Currents: Electrification and Power Politics in Bulawayo, Colonial Zimbabwe, 1894–1939." *Journal of Southern African Studies* 33, no. 2 (2007): 287–306. jstor.org/stable/25065197.

Cohn, Julie A. *The Grid: Biography of an American Technology.* Cambridge: The MIT Press, 2017.

Essig, Mark. *Edison and the Electric Chair: A Story of Light and Death.* New York: Walker and Company, 2003.

Jonnes, Jill. *Empires of Light: Edison, Tesla, Westinghouse, and the Race to Electrify the World.* New York: Random House, 2003.

Morales, Ed. "Privatizing Puerto Rico." *The Nation,* Dec. 1, 2020.

Munson, Richard. *Tesla: Inventor of the Modern.* New York: W. W. Norton & Co, 2018.

National Power Survey, 1964. Federal Power Commission.

Needham, Andrew. *Power Lines: Phoenix and the Making of the Modern Southwest.* Princeton: Princeton University Press, 2014.

Nye, David E. *Electrifying America: Social Meanings of a New Technology, 1880–1940.* Cambridge: The MIT Press, 1990.

Popovich, Nadja, and Brad Plumer. "How Does Your State Make Electricity?" *New York Times,* Oct. 28, 2020.

Rhodes, Richard. *Energy: A Human History.* New York: Simon and Schuster, 2018.

Roach, Craig R. *Simply Electrifying: The Technology That Transformed the World, from Benjamin Franklin to Elon Musk.* Dallas: Benbella Books, 2017.

Rudolph, Richard, and Scott Ridley. *Power Struggle: The Hundred-Year War over Electricity.* New York: Harper and Row, 1986.

Schiffer, Michael Brian. *Draw the Lightning Down: Benjamin Franklin and Electrical Technology in the Age of Enlightenment.* Berkeley: University of California Press, 2013.

Shamir, Ronen. *Current Flow: The Electrification of Palestine.* Palo Alto: Stanford University Press, 2013.

Thompson, William L. *Living on the Grid: The Fundamentals of the North American Electric Grids in Simple Language.* Bloomington, IN: iUniverse, 2016.

Waterworks

Allen, David, and Catherine Watling. *H_2O: The Molecule That Made Us.* WGBH Boston and Passion Planet Ltd., 2020.

Arax, Mark. *The Dreamt Land: Chasing Water and Dust Across California.* New York: Alfred A. Knopf, 2019.

Ball, Phillip. *The Water Kingdom: A Secret History of China.* Chicago: University of Chicago Press, 2017.

Burton, Virginia Lee. *Life Story: The Story of Life on Earth from Its Beginnings Up to Now.* Boston: Houghton Mifflin Harcourt, 1962, 1990.

Fishman, Charles. *The Big Thirst: The Secret Life and Turbulent Future of Water.* New York: Free Press, 2012.

Fox, Coleen A., Nicholas James Reo, Dale A. Turner, JoAnne Cook, Frank Dituri, Brett Fessell, James Jenkins, Aimee Johnson, Terina M. Rakena, Chris Riley, Ashleigh Turner, Julian Williams, and Mark Wilson. "'The River Is Us; the River Is in Our Veins': Re-defining River Restoration in Three Indigenous Communities." *Sustainability Science* 11, no. 3, May 2016.

Glennon, Robert. *Unquenchable: America's Water Crisis and What to Do About It.* Washington DC: Island Press, 2009.

Halliday, Stephen. *An Underground Guide to Sewers, or: Down, Through & Out in Paris, London, New York &c.* Cambridge: The MIT Press, 2019.

Kimmelman, Michael. "When Manhattan Was Mannahatta: A Stroll Through the Centuries." *New York Times,* May 13, 2020.

Klein, Naomi, and Rebecca Stefoff. *How to Change Everything: A Young Human's Guide to Protecting the Planet and Each Other.* New York: Simon and Schuster, 2021.

Lautman, Victoria. *The Vanishing Stepwells of India.* London: Merrell Publishers, 2017.

Miller, Todd. *Storming the Wall: Climate Change, Migration and Homeland Security.* San Francisco: City Lights Books, 2017.

Mithen, Steven. *Thirst: Water and Power in the Ancient World.* Cambridge: Harvard University Press, 2012.

Reisner, Mac. *Cadillac Desert: The American West and Its Disappearing Water.* Penguin Books, 1986, 2017.

Rutherford, Gabriella. "We Are Nature's Best Guardians, Not the State," intercontinentalcry.org, Aug. 20, 2019. intercontinentalcry.org/we-are-natures-best-guardians-not-the-state.

Salzman, James. *Drinking Water: A History.* New York: Overlook Duckworth, 2012, 2017.

Sedlack, Dave. *Water 4.0: The Past, Present and Future of the World's Most Vital Resource.* New Haven: Yale University Press, 2014.

Smithsonian. "Travel Through Deep Time With This Interactive Earth." Web application. smithsonianmag.com/science-nature/travel-through-deep-time-interactive-earth-180952886.

Sneddon, Christopher. *Concrete Revolution: Large Dams, Cold War Geopolitics, and the US Bureau of Reclamation.* Chicago: University of Chicago Press, 2015.

Solomon, Steven. *Water: The Epic Struggle for Wealth, Power, and Civilization.* New York: Harper Perennial, 2010.

Taylor, Dorceta E. *Toxic Communities: Environmental Racism, Industrial Pollution, and Residential Mobility.* New York: New York University Press, 2014.

Thompkins, Christopher R. *The Croton Dams and Aqueducts.* Charleston: Arcadia, 2000.

General

Acker, Emma, Sue Canterbury, Adrian Daub, and Lauren Palmor. *Cult of the Machine: Precisionism and American Art.* New Haven: Yale University Press, 2018.

Ascher, Kate. *The Works: Anatomy of a City.* New York: Penguin Books, 2005.

Gilio-Whitaker, Dina. *As Long as Grass Grows: The Indigenous Fight for Environmental Justice from Colonization to Standing Rock.* Boston: Beacon Press, 2019.

Hayes, Brian. *Infrastructure: A Guide to the Industrial Landscape.* New York: W. W. Norton, 2005, 2014.

Huller, Scott. *On the Grid: A Plot of Land, an Average Neighborhood, and the Systems That Make Our World Work.* New York: Rodale, 2010.

Kingfisher Visual Factfinder, The. New York: Kingfisher, 1993, 1996.

Macaulay, David. *City.* Boston: Houghton Mifflin Harcourt, 1974.

Macaulay, David. *Mill.* Boston: Houghton Mifflin Harcourt, 1983.

Macaulay, David. *The Underground.* Boston: Houghton Mifflin Harcourt, 1976.

Macaulay, David. *The Way Things Work Now: From Levers to Lasers, Windmills to Wi-Fi, A Visual Guide to the World of Machines.* Boston: Houghton Mifflin Harcourt, 1988, 2016.

About the Author

Photo by Hannah Cohen

Dan began drawing comics after taking cartooning classes at the Worcester Art Museum as a tiny human. He frequently tried to submit comics for class assignments, from elementary school all the way through college, where he studied political science, journalism, and art at the University of Massachusetts Amherst.

At the Center for Cartoon Studies (CCS) in White River Junction, Vermont, Dan began using comics to look at hidden systems for his thesis project, which forms the first part of this book.

Dan was the lead cartoonist on a comic called *This Is What Democracy Looks Like: A Graphic Guide to Governance,* which was published by CCS and distributed for free around the country. He also loves independent publishing and has self-published a range of minicomics.

Aside from comics, Dan works as an illustrator, making comics and art for investigative reports and journalism organizations like *The Nib,* WBUR, NJ Advance Media, and *Spotlight PA.* Dan is also an educator and teaches classes on making comics and comics history.

Dan lives in Vermont with his partner, Daryl, and their adorable and ferocious cat, Zuko.

See what he's up to now at dannott.com.